Artificial Intelligence Versus Natural Intelligence

Roger Penrose ·
Emanuele Severino ·
Fabio Scardigli ·
Ines Testoni ·
Giuseppe Vitiello ·
Giacomo Mauro D'Ariano ·
Federico Faggin
Authors

Fabio Scardigli
Editor

Artificial Intelligence Versus Natural Intelligence

Authors

Roger Penrose
Mathematical Institute
Oxford University
Oxford, UK

Fabio Scardigli
Dipartimento di Matematica
Politecnico di Milano
Milan, Italy

Giuseppe Vitiello
Dipartimento di Fisica
Università di Salerno
Fisciano, Italy

Federico Faggin
Los Altos Hills, CA, USA

Emanuele Severino
Brescia, Italy

Ines Testoni
FISPPA, Section Applied Psychology
University of Padua
Padova, Italy

Giacomo Mauro D'Ariano
Dipartimento di Fisica
Università di Pavia
Pavia, Italy

Editor
Fabio Scardigli
Dipartimento di Matematica
Politecnico di Milano
Milan, Italy

ISBN 978-3-030-85479-9   ISBN 978-3-030-85480-5   (eBook)
https://doi.org/10.1007/978-3-030-85480-5

Language rights for Italian edition excluded.

© The Editor(s) (if applicable) and The Author(s), under exclusive license to Springer Nature Switzerland AG 2022
The Chapter Hard Problem and Free Will: An Information-Theoretical Approach is licensed under the terms of the Creative Commons Attribution 4.0 International License (http://creativecommons.org/licenses/by/4.0/). For further details see license information in the chapter.
This work is subject to copyright. All rights are solely and exclusively licensed by the Publisher, whether the whole or part of the material is concerned, specifically the rights of reprinting, reuse of illustrations, recitation, broadcasting, reproduction on microfilms or in any other physical way, and transmission or information storage and retrieval, electronic adaptation, computer software, or by similar or dissimilar methodology now known or hereafter developed.
The use of general descriptive names, registered names, trademarks, service marks, etc. in this publication does not imply, even in the absence of a specific statement, that such names are exempt from the relevant protective laws and regulations and therefore free for general use.
The publisher, the authors and the editors are safe to assume that the advice and information in this book are believed to be true and accurate at the date of publication. Neither the publisher nor the authors or the editors give a warranty, expressed or implied, with respect to the material contained herein or for any errors or omissions that may have been made. The publisher remains neutral with regard to jurisdictional claims in published maps and institutional affiliations.

This Springer imprint is published by the registered company Springer Nature Switzerland AG
The registered company address is: Gewerbestrasse 11, 6330 Cham, Switzerland

Contents

Introduction   1
Fabio Scardigli

A Dialogue on Artificial Intelligence Versus Natural Intelligence   27
Roger Penrose and Emanuele Severino

The Death of the Emperor's Mind from an Eternalist Perspective   71
Ines Testoni

The Brain Is not a Stupid Star   107
Giuseppe Vitiello

Hard Problem and Free Will: An Information-Theoretical Approach   145
Giacomo Mauro D'Ariano and Federico Faggin

Introduction

Fabio Scardigli

This book contains the transcriptions of the talks and the debate between Roger Penrose and Emanuele Severino that took place during the conference "Artificial Intelligence vs Natural Intelligence", held in Milano, at the Cariplo Congress Center, on May 12, 2018.

Besides the keynote speeches of Penrose and Severino, there were the illuminating talks of Giuseppe Vitiello (theoretical physicist), Mauro D'Ariano (theoretical physicist), and Ines Testoni (psychologist), which gave rise to the three essays completing this book.

The conference was conceived and organized (like the previous one on "Determinism and Free Will") by a group of friends and colleagues: Fabio Scardigli, Marcello

---

F. Scardigli (✉)
Department of Mathematics, Polytechnic of Milan, Milan, Italy
e-mail: fabio@phys.ntu.edu.tw

Esposito, and Marco Dotti. Our warmest thanks go to our colleague Massimo Blasone for his help during the workshop's days.

The success of the conference was somehow astonishing, even greater than that of the previous meeting. More than 600 people crowded into the main hall of the Cariplo Congress Center and into two adjacent rooms, equipped with closed circuit television. This vividly testifies to the great interest that the general public has for the themes of Artificial Intelligence, Theory of Consciousness, Intelligent devices, and all that.

1 Understanding and Algorithms

Regarding those topics usually grouped under the heading "AI" (Artificial Intelligence), the perspectives of the two main speakers, the mathematical physicist Roger Penrose and the philosopher Emanuele Severino, are obviously quite different. Nevertheless, as the reader will soon discover, both agree that we do not yet have "intelligent devices", and also that, if we follow the vision of the so-called Strong AI (presently still the mainstream), we will never be able to build such devices. This opinion is also supported by the authors of the other essays in the book, although with their own slant.

In his main talk, Penrose focuses on the relations between the words "intelligence", "understanding", and "consciousness". "Being a mathematician", he kids, "connections among words are more important to me than their "true" meaning". Relations among concepts are more illuminating than substantial definitions, in other words. Therefore Penrose starts from the idea that the word "intelligence", at least in the standard usage, implies "understanding", and "understanding" requires some "awareness" or "consciousness". Going through different examples, discussed in great detail, Penrose shows that the machines, or software, at our disposal today are computational devices, maybe very sophisticated, but all essentially based on the ideal prototype of the Turing Machine. "Intelligence" and "understanding", on the other hand, seem to exhibit properties that escape simple computability. Intelligence is something more than mere computational ability. Examples taken from chess, mathematical induction (the intriguing Goodstein theorem), the tiling of the Euclidean plane (polyominoes, which cannot be produced through computable algorithms), all converge to show that "understanding is something which is not achieved by rules". So, a general quality of understanding seems to be that it is not an algorithm. Hence, "understanding", according to Penrose, is not an element or a result of a (very) complicated application of a set of rules (algorithm).

2 Quantum Mechanics and Consciousness

Penrose then reviews the two main theoretical pillars of modern physics, namely General Relativity and Quantum Mechanics, and points out that there is only one specific element of modern physics that cannot be reproduced on a computer, because it is not computable. To be precise, it is the measurement process in Quantum Mechanics, the so called "collapse of the wave function". This is not described by the Schroedinger equation, and cannot be implemented (not even in principle) by any computable algorithm.

In the words of Penrose "The idea is that the collapse process is something which is not computable. In fact it is something which is a bit like what free will might be. Because according to the current physics, it is making its own decisions. Somehow it decides to be here or there."

Thus, in the whole landscape of physical theories and phenomena, there seem to be only two things characterized by their intrinsic non-algorithmic, non-computable nature: the measurement process in quantum mechanics (or the collapse of the wave function), and the phenomenon of "understanding" or "awareness" proper to (human) consciousness. In his work, Penrose has been trying to connect these two concepts from many years now. Starting in particular with the book "The Emperor's New Mind", he formulates a theory according to which the origin of the non-algorithmic processes of "understanding", "awareness", and "consciousness" should be sought in certain quantum processes occurring in specific regions of the brain. Thanks to his collaboration with the anesthesiologist Stuart Hameroff, during the 1990s, some promising candidates to host these so-called "Spontaneous Orchestrated Reductions" of the quantum states were identified in the "microtubules". Microtubules are very small sub-cellular structures, located inside neurons, in the axons and dendrites.

In these structures, quantum coherence could be maintained long enough to allow the quantum collapse of a wave function to directly influence "elements", or "atoms", of quantum consciousness, named "protoconsciousness". According to Hameroff and Penrose, these could be the building blocks out of which consciousness is constructed. Obviously, in these building blocks, there is not yet a conscious aim, not yet a purpose, and even less a meaning, but out of them structures containing conscious behavior might emerge.

3 Protoconsciousness

An immediate consequence of this theory is that consciousness, far from being a purely human characteristic, should instead emerge wherever structures like microtubules are present, hence, for example, in animals like apes, dolphins, dogs, cats, or mice. This vision, also full of ethical implications, was what sparked the discussion with Severino. Moreover, Penrose also briefly introduced the possibility of "constructing" a conscious, and therefore intelligent, device. In his perspective, of course, this could be done only by aggregating elements of protoconsciousness and providing them with the appropriate environment for an "orchestrated objective reduction" of the wave function to take place.

Quite obviously, according to Penrose, insofar as they are based on purely algorithmic capabilities, the computers of today, and most likely also those of tomorrow, are and will remain "unconscious", hence lacking in real understanding and intelligence. In particular, there is no danger that "intelligent machines" will one day be able to take over the world, and threaten or destroy the human race. On this issue, it is well known that Stephen Hawking used to have an opposite view.

4 Free Will and Singularities

In connection with the theme of free will, and in the light of the above ideas, we can attempt some further consideration. As it is well known, professional general relativists are afraid of singularities, which can appear in General Relativity. Roger Penrose won the 2020 Nobel prize in Physics just because in his seminal paper of 1965 proved a theorem which stated that, under very general conditions (in particular, without requiring any spherical symmetry), singularities are a general and inevitable prediction of General Relativity (both in the past direction, i.e. the Big Bang cosmological singularity, and in the future direction, i.e. the black holes singularities).

But why theoretical physicists don't like singularities?

The standard answer is that in a singularity the predictive ability of the theory fails completely. From a singularity literally anything could come out (or enter), in a completely unpredictable way, since physical laws crash down, by definition, in a singularity. In fact, to protect the observable universe from such a monster, Penrose proposed long ago the cosmic censorship conjecture, according to which singularities are always wrapped and hidden behind an event horizon. And therefore they cannot influence the external universe.

However, according to the above considerations of Penrose, there is also another phenomenon that presents points of "singularity", where the predictive capacity of the theory fails completely: it is the single event in quantum mechanics! We have crashes of computability at every collapse of the wave function, therefore "singularities" where the predictive ability of quantum theory fails. If for example we consider the emission of a single photon from an excited atom, well, it is not possible to predict either the instant or the direction of such emission. And analogously happens for the decay of a neutron: neither the instant nor the direction of emission of the pair neutrino-electron are known, for a single event. Quantum theory predicts only probability distributions, which, although in perfect agreement with experiment, by definition, apply to classes of events only, not to single events. Therefore, "singularities", namely points where predictability crashes, appear to be present everywhere in Quantum Theory.

5 Constructing Consciousness

Two of the themes which Emanuele Severino picks out for criticism are the "place" in which human intelligence or consciousness might be located, and the possibility of "constructing" an intelligent (or conscious) device.

Severino starts by pointing out the hypothetical-deductive character of Science. All sciences, and in particular the mathematical sciences, are based on postulates, that is, propositions that are taken for granted, from which, according to certain (logical) rules, other propositions will follow. But postulates and even deductive rules are not considered to be incontrovertible truths, even by Science itself. In particular, they are themselves "conventions". Or, to use a word to which Severino attributes a wide meaning, they are "faiths". That is, expressions of the "will to power". According to Severino "the choice between two different and competing theories is ultimately determined by the ability of one rather than the other to transform the world". Science itself recognizes that it is no longer possible to build a knowledge that no one, either gods or men, could deny. Therefore, in Severino's view, Science does not aim to find incontrovertible truth; it aims rather to have power over the world.

This objective is opposed to what Severino maintains to be the primeval, foundational goal of philosophy, the goal for which philosophy was born 2500 years ago in Greece, namely to "unveil" incontrovertible truths.

6 Mathematical Modeling

However, Severino's view that the ancient Greek mathematicians, and others right up to Galileo, were trying to arrive at incontrovertible epistemic truths ("to know theorems as

God knows them", in Galileo's words), can surely be criticized. It is in fact rather clear, especially according to recent historical reconstructions (see, e.g., Lucio Russo), that the concept of "mathematical model" was already well developed in the work of Euclid, Archimedes, and others. The free choice of postulates, the possibility of "playing around" with them in order to better model a given phenomenon, or a given set of (mathematical) propositions, are all operations very clearly stated and effectively performed by ancient mathematicians, right down to the modern fathers of the scientific revolution, Copernicus, Galileo, Newton, etc. Actually, it can be safely affirmed that not one of them really believed in the construction of absolute truths; instead they were concerned with (mathematical) models performing better than the previous ones (e.g., finding a physics better than Aristotelian physics, or an astronomy better than Ptolemaic astronomy, etc.) (See also Determinism and Free Will, Introduction, on this point).

7 Manifestation of the World

Severino then introduces the concept of "manifestation of the world", the dimension where everything happens, the dimension where everything manifests itself, the dimension from which the news of every single fact or thing is drawn: "There is no step that science can take that does not spring from the manifestation of the world." According to Severino, the "manifestation of the world" is not a thing among other things, because it contains all the past, present, and future things of the world, everything that appears.

Severino appeals also to the idea of "manifestation of the world" in his second intervention, where he strongly opposes Penrose's view of "a place (in the brain) where to look for consciousness". According to Severino, looking for a place "where consciousness resides" means, in fact, considering consciousness once again merely as a specific part of that "appearing of the world", of that "manifestation of the world", which should actually be considered as the primary form of consciousness.

Nevertheless, this idea of a "consciousness of the world" would appear to share traits with the "atoms" of proto-consciousness (common to different entities) introduced by Penrose himself, and similarities can also be found in concepts considered by the other authors in this book, Vitiello, D'Ariano, and Faggin, although with different emphases and viewed from different angles.

8 Production

The core of Severino's criticism of the idea of "artificial" intelligence, or "artificial" consciousness, focusses on the idea of "production" itself. As is customary in Severino's philosophical position, the error, the major nihilistic mistake of Western philosophy (civilization), hides within the verb "produce", in the idea of "construction", i.e., of the "production" of something. Far from being innocent, as it may appear, the concept of production, "poiesis" in Greek, hides the obvious, common-sense belief that things can be made to come out from nothing, and (if desired) to go back, to return to nothing. According to Severino, the observation that things can be created and destroyed, the belief that things oscillate between being and not being, with the associated transition from not being to being, and therefore the becoming of things, which we all consider to be the supreme obviousness, is not an observable subject, it is not the substance of an observation, but is a theory, namely only one possible interpretation of reality, among many. Behind this stubborn belief, the ghost of nihilism hides itself: the belief that things are actually nothing.

Clearly, from this point of view, the idea of producing an artificial intelligence is just another manifestation of the deep nihilistic vision pervading Western civilization.

9 Artificial Intelligence

But Severino goes further. He observes that the Platonic definition of "production" as "the cause that makes a thing to go from not being to being" permeates the whole of Western thought, in philosophy, economics, law, and the mathematical sciences, and is always considered as obvious. Moreover, Western thought views Man as the only being able to organize means for the production of purposes. But, as Severino notes, this is also the definition of the machine! Except that, for the moment, machines have no purposes in their sights: "Man is that machine which organizes means in view of the production of purposes, having in mind precisely the presence of purposes, the ideal presence". In this way Severino reaches his provocative conclusion that the "natural" Man is thought of as a machine, and indeed that the world itself is thought of as a mechanism in which means are organized in view of the production of goals. And therefore, given the way in which Man has been understood in the West, then Man, or rather the natural intelligence of Man, "is" already an "artificial" intelligence.

In opposition to that, Severino reaffirms that "the manifestation of the world as a whole—this primary and fundamental form of consciousness—cannot be an object of production, at least for this reason: because the producer, if he had to produce the totality of the manifestation of the world, would have to be outside the manifestation of the world, and therefore it would be something unknown."

10 Debate

The second part of the dialog contains the interaction, or debate, between Penrose and Severino, and between the public and the speakers. Many questions involved the issue of microtubules. Regarding their capacity to survive death, and hence preserve some memories of the past life, Penrose is definitely skeptical: "I think that microtubules will not survive death any more than neurons". A second set of questions concerned tests of consciousness, and in particular ways to investigate how general anesthetics act on consciousness. The general opinion is that microtubules are very much involved in the actions of general anesthetics. Penrose also introduces the role of the cerebellum here, which controls all the "automatic" actions of a human being, and whose action seem to be entirely unconscious, as opposed to the role of the cerebrum, where conscious actions seem to play an explicit role. As confirmation of his ideas, Penrose points out that microtubules are abundant in pyramidal cells, which are in turn abundant in the cerebrum, while on the contrary they are not found in the cerebellum. What is interesting here is that consciousness seems to appear in pyramidal cells. Other tests of conscious understanding are carried out with specific chess problems, designed for humans and for computers. Here one sees the distinction between "understanding", e.g., what a barrier of pawns does, and the purely "mechanical" calculation which a computer performs. These tests clearly illustrate the difference between conscious thinking (or conscious understanding), and mere computation.

11 Consciousness in Animals

On the contrary, according to Penrose, the notion of creativity is misleading and ambiguous: creativity is not a good test for consciousness. Understanding is something where you can see the difference between conscious or unconscious action. Instead, it is very hard to see the difference between creativity and just random production of something (which is considered different from what has been "created" before). So, creativity, as opposed to understanding, is extremely hard to test, and it is difficult to be objective about that. Finally, Penrose supports the idea that the phenomenon of consciousness is present also in animals. This is consistent with his vision of microtubules as the arena in which protoconscious events, and finally consciousness itself, draw their origin. Microtubules are very present in many "superior" animals, and so therefore should be consciousness, in particular in dogs, cows, elephants, monkeys, gorillas, dolphins, mice, etc. From this point of view, ethical conclusions could be drawn, such as the respect we owe, not just to other humans, but also to (many) other creatures.

12 Consciousness and Language

The idea proposed by Penrose and Hameroff (consciousness emerges out of a cumulative process of elementary "protoconsciousness" acts, until we finally arrive at the wonders of human mind), calls for a disturbing observation about the "strong" AI program. As is well known, and as has happened historically, such a program starts from (formal) languages and aims to rebuild "intelligence" through software (i.e., computer programs based on languages), following a kind of top-down process. And this is done with the fairly overt conviction that perhaps from the construction of intelligence we can then pass, by continuing on the same path, to the construction of consciousness.

However, according to the ideas put forward by Penrose, Nature seems to arrive first at the construction of simple forms of consciousness, which actually appears to be a fairly common phenomenon at least in the "higher" animals. Only after that, Nature builds languages (even complicated ones); and languages, at least the advanced ones, seem to be relevant only for a single species, humans.

From this point of view, the Strong AI program looks truly "artificial", in the sense that it is moving along a path opposite to the one followed by natural evolution. Humans are trying to build consciousness starting from language, whereas Nature has built languages starting from (proto)consciousness!

On the other hand, as D. Hofstadter once said, perhaps Artificial Intelligence should be compared to a modern jet airplane: high-performance when it comes to specific tasks, but generally unable to do things that a sparrow, which in the metaphor represents human intelligence, can easily do. A jet can fly from London to Milano in a hour, an impossible task for a sparrow or a pigeon. But try to land a jet on a gutter…

13 A Place for Consciousness?

Severino's counter reply to Penrose well illustrates the distance and the differences between their approaches. Severino openly criticizes Penrose for ignoring his words about the "manifestation of the world as a primary form of consciousness", although this concept is present, according to Severino, in many important exponents of contemporary culture, such as Descartes, Kant, Brouwer. When Penrose repeatedly looks for the "place where consciousness resides", he simply shows, in Severino's view, that he still imagines consciousness as a thing among things, namely only as a part within that "manifestation of the world" which is actually the primary form of consciousness.

A final topic where the disagreement is particularly clear is the practical nature of science, which according to Severino means this: the conceptual articulations of scientific knowledge allow a power over the world superior to other conceptual articulations, such as the conceptual articulation of the alliance with the sacred, i.e., prayer. So, the conceptual articulation of modern scientific knowledge is formidable! It is what today allows the greatest power over the world. But power is one thing, truth is another. Severino comes back to his initial argument about the technological power of scientific theories. His conviction is that a scientific theory ultimately is evaluated on the basis of its technical capacity to transform the world, not its ability to truly represent it, or effectively explain it. The theme of intersubjectivity in science is recalled here by Severino. According to Popper, for there to be power, intersubjective recognition of the power in question is needed, and this means that others have to clearly perceive the transformation of the world.

14 Science and Technology

On this last point Penrose replies forcefully (and in my opinion correctly), by reaffirming the traditional separation between science and technology. Deep ethical issues enter the discussion at this point. In Penrose's words: "My way of looking at science is to try find out what is true about the world. So, there is no moral issue involved. I mean, the moral issue is a separate question." Science tries to understand the way the world operates, how it works. Then, there is technology, which works on how to use scientific knowledge. Technology has a close relationship with science, but it is not science. Technology has a huge and continued impact on what people do in their everyday lives. And this of course raises the moral issues. So, according to Penrose, the use of technology, and in particular the good or bad use of science, is deeply entangled with morality. Penrose is clear about that: "When I'm doing science, I'm trying to develop an understanding of the ways world operates, and I am not looking for power". Science is not trying to control the world. That is the aim of technology. Technology and morality are areas clearly separated from science, although they depend on science; when science changes, these two other areas have to pay respect to science, and see what it tells them.

Finally, Penrose points out once again that his and Hameroff's ideas would tend to consider consciousness as a phenomenon that is not restricted to human beings, but appears also in other animals. This implies that a moral issue is involved in how we deal with animals, and this is a further example of the way science influences our moral beliefs.

15 Chatbots

The essay by Ines Testoni, a former disciple of Severino, presents a number of interesting points, of which I can only discuss a few here. The essay opens by presenting an experiment, actually performed in 2017 at the Facebook Artificial Intelligence Research (FAIR) group. Two computer programs (software) were trained in English conversations, and then allowed to chat autonomously with each other using the English language, also in a non-human way. The two chatbots seemed to progressively invent a language inaccessible to humans. The dialogue established between them could even be interpreted as the constitution of an autonomous consciousness in computers. Perhaps—Testoni provokes—artificial intelligence systems are anticipating, or even realizing, the formation of a "quantum Turing machine", which humans do not yet know how to construct.

This interpretation would agree perfectly with the visions of Putnam [17] and Chalmers [18], who hold that the material constitution of the mind is completely irrelevant to the production of thought and consciousness.

In other words, mental properties are organizationally invariant, in the sense that the material support may change. If any mental state is organizationally invariant, then when the brain dies, it would theoretically be conceivable to replace the grey matter with an AI system. This intriguing prospect has also been explored by D. Hofstadter and D. Dennett, in their famous book "The Mind's I", and, in a different context, by W. Pfister in the beautiful and unsettling movie "Transcendence".

16 Quantum Turing Machines

As we know, Roger Penrose deeply disagrees with the "strong AI" thesis: AI systems work only through formal language, while human thought is characterized by a functioning that cannot be reduced to computational processes. If the core of (human) consciousness is the set of non-computable, non-algorithmic quantum collapse processes taking place in microtubules, then this implies that (human) consciousness is not representable by any conventional Turing machine, and that the human mind has abilities that no AI system could possess, because of the non-computable physics involved in the OrchOR mechanism [19].

The only mechanism which, at least in principle, can escape this state of affairs, and so have a chance to "exhibit" consciousness is a Turing machine itself based on quantum physics, i.e., a quantum Turing machine. From the above, Testoni puts forward an intriguing thesis: quantum mechanics could be the common element between organic and inorganic matter that causes mind and consciousness. If so, we cannot say that consciousness is only human, just because of quantum mechanics. On the contrary, consciousness may in principle appear in both organic and inorganic matter precisely because quantum mechanics is at the base of all known forms of matter.

In her conclusions Testoni points out that, according to Severino, "consciousness is the phenomenological manifestation of everything, which is eternal like any matter and is identical to itself, and so cannot be reduced to matter. The relationships between consciousness and matter (grey matter in the brain, or other) cannot be reduced to their reciprocal identity." If consciousness (the manifestation) is not restricted to a brain cavity, then we must recognize that it transcends the individual and human dimension and, thus, humans' ability to recognize its presence. Here there emerges a connection between Severino's idea of a general consciousness as "manifestation of the world", and the Penrose-Hameroff concept of "elements of protoconsciousness", which should be present wherever a wave function is collapsing.

17 Machine Making Mistakes

The essay by Giuseppe Vitiello illustrates the most significant aspects of a Dissipative Quantum (Field Theoretic) model of the brain (and consciousness). Already from the title, Vitiello recalls the fundamental role played by (dissipative) chaos for the ability of the brain to respond flexibly to the outside world. Freeman stressed this concept, and the remark attributed to Aristotle, that "the brain is not a stupid star", vividly depicts the fact that the brain in its perennial trajectory never passes through the same point in a fully predictable way. A brain behaves like a 'machine making mistakes', that is, as an intrinsically erratic device.

"Coherence" is a central concept in Vitiello's approach to unveiling the marvels of the (human) brain. Observationally, neural activity in the neocortex displays the formation of extended configurations of oscillatory motions. These configurations extend over regions with linear dimensions up to twenty centimeters in the human brain and have almost zero phase dispersion. The presence of some sort of "cooperation" over such large regions implies that brain functions cannot be explained using only the knowledge of individual elementary components. Brain activity demands the introduction of the notion of "coherence": a widespread cooperation between a huge number of neurons over vast brain areas. The natural mathematical tool to describe coherence is Quantum Field Theory (QFT), since its formalism has proved to be very useful in the study of biological systems in general and of the brain in particular. The mathematical formalism describing "coherence" gives a well-defined meaning to the notion of the emergence of a macroscopic property out of a microscopic dynamic process. The macroscopic system possesses physical properties that are not found at the microscopic level.

Vitiello stresses that the use of QFT rather than quantum mechanics (QM) to describe the brain is necessary because QFT allows the description of the different phases in which the system may be found. Technically, this happens because infinitely many unitarily inequivalent representations of the canonical commutation rules (CCR) exist in QFT, while, on the contrary, QM allows only unitarily (and therefore physically) equivalent descriptions (for systems with a finite number of degrees of freedom). Systems that may have different physical phases, like the brain, must be described by QFT, which may account for the multiplicity of their phases and the transitions among them, something that QM is not equipped to do.

18 Dissipative Quantum Brain

Another fundamental observation by Vitiello is that the brain is an open system, in full interaction with the environment. This structural openness has led Vitiello and others to the formulation of a dissipative quantum model of brain.

A key aspect of the model is the identification of the acquisition of a "new" specific memory, with a specific fundamental state (among the infinitely many, unitarily inequivalent states), also known as the "vacuum", to which the brain-environment system has access. This is the "secret" of memory, according to the quantum dissipative model of the brain. This set of memory states, i.e., set of vacua, can be described as a "landscape of attractors". Thoughts are in principle conceived as having chaotic trajectories in this landscape of vacua. Each act of recognition, of association with a specific memory, can be depicted as the approach toward a nearby attractor, and the consequent capture by it. According to Vitiello, this represents an act of intuitive knowledge, the recognition of a collective coherent mode, "non-computational" in nature and not translatable into the logical framework of a language. The ultimate "non-computational" nature of the thought re-emerges here once again, as in the views of Roger Penrose.

The trajectories in the landscape of attractors, from memory to memory, are classical chaotic trajectories, although they "connect" quantum states; they are not periodic (a trajectory never intersects itself), and trajectories that have different initial conditions never intersect; instead they diverge (exponentially). Here, Freeman's initial intuition finds a rigorous expression: chaotic trajectories originate the ability of the brain to respond flexibly to the outside world and to generate novel activity patterns, including those that are experienced as fresh ideas. This wandering is a characteristic trait of brain activity, of thinking. This is why the brain is not a stupid star. It behaves rather as an "erratic device", a "mistake-making machine".

The non-computational activity of an erratic device should be, according to Vitiello, the hallmark of an intelligent device: exactly what Penrose argues to be the distinctive character of consciousness.

Of course, the idea of a conscious agent as an intrinsically erratic device, a mistake-making machine, goes directly against the standard program of (strong) AI; which actually tends to produce exactly the opposite, i.e., obedient, predictable, loyal machines, possibly useful to improve our limited abilities. Essentially, AI projects today are still limited to designing "stupid stars".

19 Inner Experience

The aim of the essay by Mauro D'Ariano and Federico Faggin is very ambitious: the authors present the main lines of a quantum information theory of the concept of "inner experience". Essentially they propose a coherent theory which solves what D. Chalmers calls "the hard problem of consciousness", namely the origin and the properties of that "inner experience" which is experienced by each of us daily. The origin of qualia (=feelings, sensations), and of self, thus finds a natural place in the theoretical scheme discussed here.

Some of the starting points of this essay have also been partially described in previous publications or talks by Faggin and D'Ariano.

For example, the authors consider, as Penrose does, that "true" intelligence requires consciousness, something that our digital machines do not have, and never will. These authors are also opposed, like Penrose, to the standard AI view of human beings as a kind of "wetware". They contrast the strong AI belief that consciousness emerges from brains alone, as a product of something similar to the software of our computers, as well the physicalist view that consciousness "emerges from functioning", like some biological property of life.

On the contrary, the authors hold that the essential property of consciousness is the ability, the capacity, to feel. Of course, the ability to feel implies the existence of a subject who feels, a self. Therefore consciousness is inextricably entangled with a self which (or who) feels "inner experiences". Central to the discussion is therefore the construction of a theory of "qualia", and the solution of the "hard problem of consciousness", namely to explain the existence and dynamics of qualia.

Furthermore, the authors assume the point of view put forward by D.Chalmers, who argues that consciousness is a fundamental property, ontologically independent of any known (or even possible) physical property. All information-bearing systems may be conscious; consciousness is an irreducible aspect or property of nature, not an epiphenomenon. It is not emergent. To some degree, consciousness is an aspect of reality ab initio. This vision leads the authors (like Chalmers) to a qualified panpsychism. Here similarities clearly emerge with the Penrose-Hameroff idea of "atoms" of protoconsciousness, which should be present wherever there is a collapsing wave function in the Universe (not to mention the Severinian "manifestation of the world").

20 Quantum Panpsychism

To solve the hard problem of consciousness, namely the issue of explaining our experience, sensorial, bodily, mental, and emotional, including any stream of thoughts, the central proposal of D'Ariano and Faggin is panpsychism, with consciousness as a fundamental feature of "information", and physics supervening on information. The hypothesis of the theory presented here is that a fundamental property of "information" is its "being experienced" by the supporting "system".

The information involved in consciousness should necessarily be quantum for two reasons: its intrinsic privacy and its power of building up thoughts by entangling qualia states. Authors call their view "quantum-information-based panpsychism".

Following Chalmers, they hold that: "Consciousness is a fundamental entity, not explained in terms of anything simpler". But they go further, realizing the generic assertions of Chalmers in an explicit technical model. Precisely, they postulate: "Consciousness is the information-system's experience of its own information state and processing." Moreover, they postulate the quantumness of experience: the information theory of consciousness is a quantum theory. And again, they state the qualia principle: "Experience is made of structured qualia". Qualia (phenomenal qualitative properties) are the building blocks of conscious experience.

21   Choice as a Quantum Variable

Particularly intriguing is the discussion of the concept of Free Will. While consciousness is identified with quantum information, free will produces public effects which are classical manifestations of quantum information. Will is said to be free "if its unpredictability by an external observer cannot be interpreted in terms of lack of knowledge". The authors argue that if "choice" is a random variable, then it cannot be a classical one, since it could be always interpreted as a lack of knowledge. Therefore the free will of choice should be a quantum variable, whose randomness cannot ever be interpreted as a lack of knowledge. Here the connection with the Free Will Theorem due to Conway and Kochen is clear. Since choice is a quantum variable, its randomness should be of pure quantum origin. The randomness, or the freedom, in the choices of consciousness has the same quantum nature as the randomness of a quantum particle. The electron (or whatever other particle) is "free" to choose which slit to cross: in fact we cannot predict its individual path, and this impossibility is not due to a lack of knowledge. This is exactly what is stated in the Free Will Theorem: the degree of free will of the electron is the same as the degree of free will of the external observer! In general the evolution of consciousness is quantum, so its kind of randomness is quantum and cannot be interpreted as lack of knowledge (no hidden variables!); as such, it is therefore free.

22 Free Will and Consciousness

The D'Ariano-Faggin model describes several further aspects of conscious behavior, using also the rather specialized techniques of quantum circuits. In particular, memory is considered to be essentially classical, and the transfer of quantum experience to classical memory, and conversely, of classical memory to quantum experience, are detailed with rigorous technicalities. In their approach, free will and consciousness are deeply connected, allowing a system to act on the basis of its qualia experience by converting quantum to classical information, and thus giving causal power to subjectivity. The authors believe that physics describes an open-ended future, because the free will choices of conscious agents have yet to be made. We, assert the authors, as conscious beings, are the co-creators of our physical world. Here we can glimpse an idealistic perspective, partially connected with the "manifestation of the world" described by Severino.

Of course, the idea that our physical world is created through our free-will choices depicts the world as a series of purely creative acts, based on the appearance from nothing of the elementary (quantum) event, and on its going back to nothing. The perspective of D'Ariano and Faggin is, from this point of view, radically opposed to Severino's. We can thus conclude by saying that these authors fully embrace the vision of Heraclitus, against that of Parmenides and Severino.

References:

1. Penrose, R. (1989). *The emperor's new mind*. Oxford: Oxford University Press.
2. Penrose, R. (1994). *Shadows of the Mind*. Oxford: Oxford University Press.

3. Penrose, R. (1997). *The large, the small, and the human mind*. Cambridge: Cambridge University Press.
4. Penrose, R. (2004). *The road to reality*. London: Jonathan Cape, Penguin RH.
5. Severino, E. (2016). *The essence of nihilism*. London: Verso Books.
6. Severino, E. (1979). *Legge e Caso*. Milano: Adelphi Edizioni.
7. Severino, E. (1981). *La struttura originaria*. Milano: Adelphi Edizioni.
8. Russo, L. (2004). *The forgotten revolution*. Berlin: Springer.
9. Conway, J., & Kochen, S. (2006). The free will theorem. *Foundations of Physics, 36*, 1441.
10. Bell, J. S. (1987). *Speakable and unspeakable in quantum mechanics*. Cambridge: Cambridge University Press.
11. Scardigli, F. (2007). A quantum-like description of the planetary systems. *Foundations of Physics, 37*, 1278.
12. Scardigli, F., 't Hooft, G., Severino, E, & Coda, P. (2019). *Determinism and free will*. Heidelberg, Springer Nature.
13. Hofstadter, D. R., & Dennett, D. C. (1982). *The mind's I*. New York: Bantam Books.
14. Vitiello, G. (2001). *My double unveiled: The dissipative quantum model of the brain*. Amsterdam: John Benjamins Publishing.
15. Faggin, F. (2019). *Silicio*. Milano: Mondadori.
16. D'Ariano, G. M., Chiribella, G., & Perinotti, P. (2017). *Quantum theory from first principles*. Cambridge: Cambridge University Press.
17. Putnam, H. (1975). *Philosophical Papers Volume 2: Mind, Language and Reality*. Cambridge University Press.
18. Chalmers, D. J. (2011). A Computational Foundation for the Study of Cognition. *Journal of Cognitive Science, 12*(4), 325–359.
19. Hameroff, S., & Penrose, R. (2014). Consciousness in the universe: A review of the 'Orch OR' theory. *Physics of Life Reviews, 11*(1), 1–40.

A Dialogue on Artificial Intelligence Versus Natural Intelligence

Roger Penrose and Emanuele Severino

1 Penrose: First Intervention

Well it's a great pleasure and an honour for me to be here, and I'm using this wonderful piece of modern technology here, which … I hope it works… [the device is an old projector for transparencies].

So this is the title I've given to this talk:

Why New Physics is needed to understand the Conscious Mind.

R. Penrose (✉)
Mathematical Institute, Oxford University, Oxford, UK

E. Severino
Brescia, Italy

Now, this talk is about artificial intelligence and real intelligence, and it is also about why I don't think we have now artificial intelligence, so I will come to that.

But let me first say: well there are many things that I'm not going to talk about, which are many features of consciousness such as feeling a pain, appreciation of the colour red, or love, or intention and all the things that are usually attributed to conscious beings.

I am only going to concentrate to one thing, which is the word "understanding". I relate three words, which I don't know the meaning of.

One of them is "intelligence", one of them is "consciousness", and the other one is "understanding" (see Fig. 1). Coming from a mathematical background I don't need to understand the words if I can say something about the connections between them.

So I would say that in ordinary usage the word "intelligence" requires "understanding", and I would say that in the normal usage "understanding" requires some "awareness". So I would say that for an entity to be "intelligent", in the ordinary usage of the word, it would have to be aware, it would have to be conscious. And I don't think the devices that we know about are conscious. But I am not directly addressing the issue of consciousness. What I will be more directly addressing is the issue of understanding. Because I believe understanding is something that we can say something quite objective about.

And I start with a chess position which I made up recently. We know that there are computer controlled chess players which play chess better than any human being. So you would say that, surely, these chess programs understand chess.

Well, the most effective of these chess programs is called Fritz. And this position (Fig. 2) was given to Fritz. A position that I made up deliberately, because if you notice the

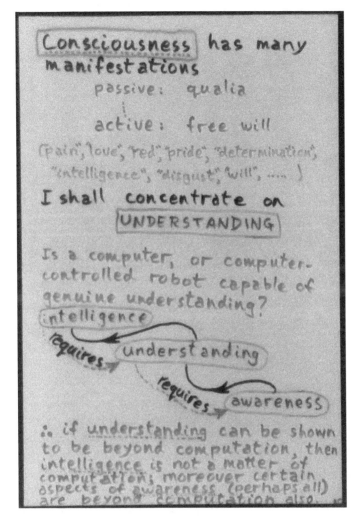

Fig. 1 Intelligence, understanding, awareness

black pieces, well there are many more black pieces than whites. You may worry that this position is not a legal position because there are three black Bishops, but it isn't an illegal position because two have come from pawn transformation.

R. Penrose and E. Severino

Fig. 2 A chess position

There's nothing illegal about the position. It's a stupid position. It would never occur in an actual game, which is one of the reasons why Fritz has no idea how to play this position. Because to a human being, who knows very much about chess, he would see that, for the black, the knight is trapped, a bishop can move around a little bit, the king cannot get out, and so these pieces are all trapped behind this barrier. The pawns also cannot move. And then there is no way to lose, there is no way black can win, it's an obvious draw.

However, you give this position to Fritz, and it presumes it is a win for black!

And that, I should say, is its downfall. Because it plays the game for a while, it makes a completely stupid move and loses the game! Which is what I expected. It loses the game because by some kind of count, that I don't know exactly what it is, it thinks, "thinks" maybe is the wrong word, it decides that it is a win for black. Probably because just counting pieces, maybe it sees that thirty moves ahead it doesn't make any difference, it's still got more pieces, and therefore it concludes that it is a win for black.

Fritz is of that opinion. Then it comes up to close to the "Thirty move rule", the rule says "if your pieces have not been taken or a pawn has not moved then it is a draw". And since Fritz considers that the game is a win for black, a draw is a disaster!

Then, either it sacrifices its bishop, or it takes the bishops away from the line here, and the king can take a pawn and then another pawn can march up and you can make three queens, completely overwhelming the position, and it is a win for white!

Eventually Fritz agrees it is a win for white, but you can see, it has no understanding whatsoever of the game of chess, because any player who knows not very much of the chess at all would see this is a draw, and it is very stupid to give up a bishop, or release that pawn which can become a queen…

Ok, that's only a special case. I just use it as a demonstration to show that, despite the training that these programs have, it doesn't seem that they understand the game, and I don't think they do. Now, the point about chess of course is that it is a finite game, so you might run out of positions like this. I'm sure that, as I speak, there are people trying to program Fritz, so that, if it might be given a position like that, it would not make that stupid move. It is not so easy. But of course you can! However, Fritz is not deciding how to change the program. It's a human being, who understands what is going on, who understands the program that underlies how Fritz behaves. So the understanding is on the part of the human being, not on Fritz's.

But nevertheless chess is a finite game, so you might worry that, ok, it could be completely analyzed, and then mistakes like that will not be made anymore.

So we have to concentrate on the infinite. Some people think that infinite is something you can't understand, we cannot contemplate. But nevertheless, this is not true. We can understand infinite. Any one of you is good at that. If you get two even numbers, and add them together, you get another even number. Well, that is a statement about all numbers.

We can do that. But that is a statement about an infinite number of things. There's no problem about talking about infinite number of things and understanding an infinite number of things. Sometimes it can be very difficult. Sometimes it can be easy.

What I claim is that the machines don't have this ability. And there are theorems which more or less tell you that. Let me give you a famous theorem. I'm going to give you the Goedel theorem.

No, wait a minute, first I want to tell you the way we learn about the infinite when we go to school.

We learn mathematical induction. This is one way we can understand the infinite. I'm talking about natural numbers, which are all the non-negative integers. Now, suppose you have a proposition, P(n), which depends on the natural numbers. Now if you prove that this proposition is true for zero, or if you can see that it is true for zero, and you can prove that if it's true for n then it's true for n + 1, then you know it's true for all numbers!

Just establishing two things, those two things, tells you that the proposition is true for all numbers. So this is a way you can establish things that are true for all numbers.

But I am going to say something about the limitation of that.

The first thing is a very general statement, which is due to Goedel.

But I shall give you it in the form due to Turing, which is a very nice way of saying it.

Suppose you're given a set of rules, I'm calling the set R, any set of rules which could be checked. You see, you have a set of propositions, each of which follows from the one before by the rules R. And I'm just insisting that a computer could check whether the rules are correctly carried out or not.

Now, let's say these rules are meant to be ways of proving theorems about numbers. It could be Fermat's last theorem which was proved by Andrew Wiles, or it could be the Goldbach conjecture which says that every even number greater than 2 is the sum of two prime numbers, still not proved. That's the kind of propositions I'm talking about.

Propositions about an infinite numbers of things.

The rules have to be such that, if you follow them correctly, namely you feed the rules a particular statement, and the rules chug away, and if they can obtain "Yes that's true", ok then you have to believe that yes it is true. They might say "no, it's false", ok, or they might not say anything. They might just go on, and on, and on, and on forever.

Now let's suppose that you believe that the rules never say a false statement, like 2 plus 2 equals 3. Let's assume that they never give you something which is blatantly false, but they only give you things that are true.

Then, what we are going to show is that you can construct a particular statement about numbers, like the Goldbach conjecture, like the Fermat theorem, a very specific statement, and you can see from the way you constructed it that:

first of all, the statement is true;

and secondly, the statement is not obtainable by means of the rules.

Now, this is a very remarkable thing. When I first heard this I was stunned. Because what it told you is not that these things are not provable, because the very thing that you construct is something that you know is true. But you know it's true only because you trust the rules. So this means that, if you believe that the rules only give you truths, then your belief transcends the rules themselves.

A very remarkable thing!

It means that the understanding of why the rules work tells you something that the rules cannot achieve. It tells you that the rules cannot achieve something, but nevertheless that something is true. You can see it is true by the way you constructed it, if you trust the rules. When I first heard about this, I was amazed!

Because it seems to tell us that there is something special in our understanding—you don't have to invent it, you don't have to be Goedel, you don't have to be Turing, you don't have to have the inspiration to see why this has to be true. All you have to do is to follow the reasoning, and you can see by following the reasoning that the statement that comes out of G(R) is true, if you trust the rules. The understanding of why the rules work gives you something that the rules don't give you. Well, this fact clearly tells you that understanding is something which is not achieved by rules.

Now there is something more about that, which I come to in a minute. But let me give you some examples of something that computers cannot do.

There are many such examples. This is a nice one because it is very easy to understand. Suppose you are given things like this (Fig. 3), which are made out of squares, all same size, and they are glued together along the edges, and you can make various shapes in the plane. They are called polyominoes.

It may be that with some of these shapes you can cover the entire Euclidean plane, just using that shape. Here is

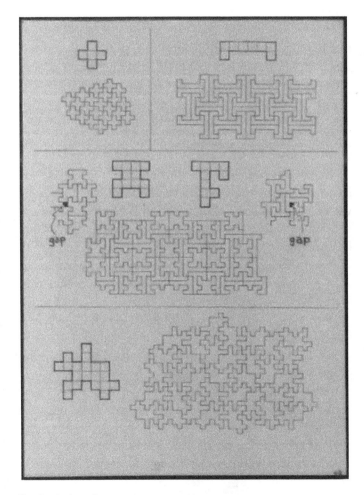

Fig. 3 Polyominoes

an example by which you can cover the entire plane. And here there is another one which leaves some gaps in-between the shapes, and so on.

Now, there is a mathematical theorem, due to Robert Berger, that states there is no computer algorithm which will tell you, "yes" or "no", whether a finite set of polyominoes will tile the plane. That's very remarkable. Let me give you an example.

Here is an example (Fig. 4), I made this one up myself but it shows you the difficulties. These three shapes will tile the plane. But if you look at that pattern, and you stare at it for a while, can you see the pattern by which it's constructed?

A very well known mathematician told me "I cannot see how this is constructed, how would you do it? Is there an algorithm?".

Well, there is an algorithm to see how to do this. You have to look in my collected works. There is one paper where you can find how to do this. But you are not allowed to look at my papers, that's cheating! That's not the kind of algorithm, or procedure, I mean!

Nevertheless, there is a way of covering the entire plane with these shapes, but I don't know any computer program able to do that!

I just gave you this as an example of something which is not computable.

Yet it's possible to understand it by using the right kind of reasoning.

Maybe I don't know whether it is true in all cases. However, it seems to be possible that you could tell with an insight, at least in many cases, whether these things can be done. But there is no general algorithm for that.

Let me give you another example which is related to this.

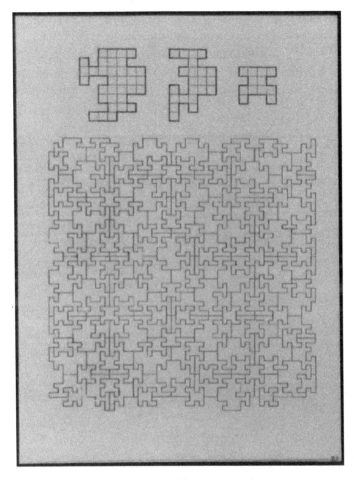

Fig. 4 Tiling the plane with polyominoes

This is an example which was put forward in 1945 by a mathematician called Reuben Goodstein, and I want to try and tell you what it is (Fig. 5).

You take any natural number, here I take as an example 1077. First you write this number in binary. You are representing this number as a sum of different parts of the binary notation. You write also the exponents in binary.

Now I'm going to give you two operations.

One operation is A. What A says is: look at all the "twos" and replace them with "threes". Ok, well defined

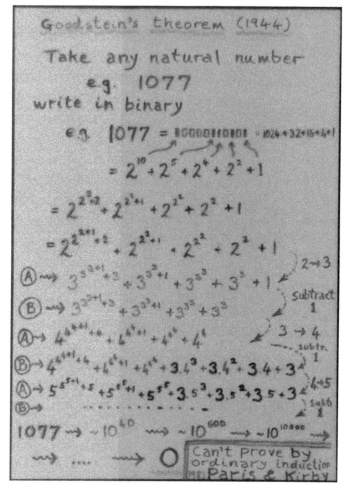

Fig. 5 Hare and tortoise

operation. The number gets much much bigger, never mind. That's operation A.

Then Operation B: subtract 1.

Ok, it comes down just a little bit.

Then again, Operation A: replace all the "threes" by "fours".

Then Operation B: subtract 1.

This is an amazing example of hare and tortoise. The tortoise always wins. That's remarkable. What happens with these numbers?

The first is 1077, the next one is about 10^{40}, the next one 10^{600}… and bigger and bigger and bigger… and finally you end up with zero! Very extraordinary! Because the number becomes huger and huger, but subtracting one (the tortoise) in the end always wins. Now this is remarkable not just because it's true, but because it was proved by J. Paris and L. Kirkby (in 1982) that you cannot prove this theorem using ordinary mathematical induction. This is an example of Goedel's theorem. To prove it you have to use the so-called transfinite induction.

I would suggest, if you want to try, try it with 3, or also try with 4.

I would not recommend that you put it on your laptop. I would not recommend that you use your mainframe computer. I would not recommend that you use any computer. I would suggest that you use a pencil and paper and you can probably convince yourself that it will eventually come down to zero. But if you put that on any computer which exists today in the world, or is conceivable in the world, it will never do it. The number gets so huge, it's remarkable… even with four… But we can understand that. It is not so obvious, but we can.

How do we know that?

How is it possible?

Some people say we have an algorithm in our head which is so complicated that we cannot apply the Goedel theorem. And that's why it doesn't work.

I don't believe that.

One reason why I don't believe that is because ... how did our understanding come about? Well, by natural selection! That I do believe!

But natural selection is not very good at making mathematicians.

Understanding in a general way, yes indeed, it was very important.

But understanding mathematics? Not really! Here is a mathematician (see Fig. 6, cartoon). Probably he has an algorithm in his head, but he is going to be eaten by the saber-toothed tiger! ... I'm trying to say that being a mathematician is not really a selective advantage!

Fig. 6 Mathematics and natural selection

So, how is it that we have an understanding? Why can we do things that no conceivable computer can do? (We have seen that with the Goodstein theorem.)

What is the general quality of understanding? Which is not an algorithm! Understanding is not an element of a very complicated algorithm. That is all I am trying to say.

So, what is the selective advantage of the general quality of being able to understand? And, this is not a computational quality. That is the message I'm trying to make.

When I was a student in Cambridge I went to a course on mathematical logic. There I learnt about Goedel and Turing, and I was amazed. But then I thought: ok, if our heads are not governed by algorithms, so that they could look like a computer, then what could it be?

I was trying to do physics. I was trying to understand how the physical world operates. And I was able to understand that you can put on a computer these very complicated things, like Einstein's equations of general relativity, and simulate things like black holes merging, and the emission of gravitational waves. People did very complicated calculations to work out the exact shape of the gravitational waves' signal which indicates that a pair of black holes is coming together. Computations! With computations you can simulate the action of Einstein's theory to a tremendous precision.

Ok, so what can we do in our heads? Maybe it's not relativity, maybe it's ordinary Newtonian physics. Well, with that you can do very, very good computations.

And what about quantum mechanics? Well, that's an interesting question.

Because you can follow the Schroedinger equation; again that can be quite difficult, but again it's purely computational.

When I was a graduate student, I went to a course given by the great physicist Paul Dirac, and in the very first lecture he had a piece of chalk which he put in two places. He was talking about the superposition principle in quantum mechanics, which says that a thing that can be in a superposition of states can be in two places at once, with certain complex numbers as weighting functions.

And then Dirac explained why a piece of chalk cannot be in two places at once, and my mind unfortunately wandered, and I never heard the explanation!:)

So I have worried about that ever since!

He said something vaguely about energy, but I don't know what that was.

So, from then on, I worried about how on Earth a piece of chalk cannot be in a superposition. Because, according to the rules of quantum mechanics, you can in principle put a piece of chalk in a superposition.

Schroedinger showed that according to quantum mechanics a cat can be dead and alive at the same time. I should point out that he worked out this example just to show that his own equation cannot be the whole story. There must be something else. He was deliberately showing the absurdity in his own equation. I think that's very remarkable!

Now I'm going to tell you about quantum mechanics. I'm using this picture (Fig. 7) to illustrate quantum mechanics. The origin of this picture is because I was invited to give a lecture by the Christian Andersen Society in Copenhagen, and this happened because I wrote a book titled The Emperor's New Mind, which is a play on The Emperor's New Clothes by Andersen.

This picture is part of the lecture I gave, and it illustrates quantum mechanics. How can it illustrate quantum mechanics?

The picture is really made of two things. The top is the classical world. You see separated things, the buildings,

Fig. 7 Classical world and quantum world

people around, they have a location, here and there, etc. The bottom is the crazy looking quantum world, which is all entangled up in complicated ways, and that's all under the sea. And what is the role of the mermaid?

Well, the mermaid is magic. She represents how to get from one world to the other. She is the magic of the reduction of the state, or the collapse of the wave function.

It is what you have to do in quantum mechanics. You follow Schroedinger's equation up to a point, and then you say: well, no, this is getting too wild... dead cats and live cats. Ah, look into the box, and your consciousness is getting involved with that, and you somehow make the cat be dead or alive instead of being in a superposition.

I never believed that. There seems to be something out there, in the physics of the world, which is making that choice, and that is the magic of the collapse of the wave function, or the reduction of the quantum state, which is what the mermaid is doing.

This is how classical physics comes out of the quantum world.

It is not coming out in the way we understand quantum mechanics now.

So I gradually began to think that this thing that I didn't understand in the Dirac lecture is really profound. Something missing in the physics.

And it is the only thing I could see that you could not put on a computer!

So it seemed to me that this must be what we can take advantage of. What consciousness is made of.

Let me now illustrate the argument by giving you an example, a formula that I elaborated after a lot of thinking (a Hungarian physicist, L. Diosi, came up with a similar formula earlier than me) (Fig. 8).

Suppose you have a body and you try to put it into two positions at once, and you want to know how long that will last. Now we say it will become one or the other in a certain length of time which depends on an energy.

I don't know what exactly Dirac was saying, but what you do is to look at the energy in this configuration and

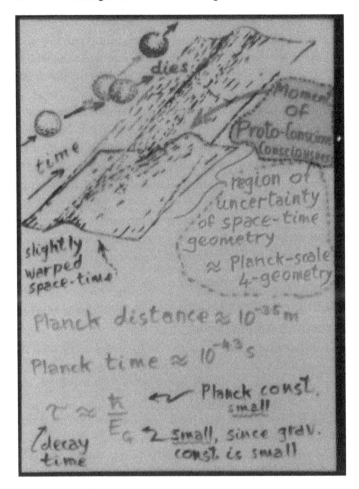

Fig. 8 Collapse of the wave function

the energy in that, and to the mass distribution, and you subtract one from the other. Then you work out the so-called gravitational self-energy of this system of two configurations, namely E_g, and finally the lifetime of that superposition of states is given by $\hbar/E_g$.

So, there is a formula that will tell you when Dirac's piece of chalk would become over here, and when it would become over there. The formula will not tell you where, but it will tell you when it becomes one or the other configuration.

The idea is that this process is something which is not computable. In fact it is something which is a bit like what free will might be. Because according to the current physics, it is making its own decisions. Somehow it decides to be here or there, somehow it decides… But I don't know what that means!

So, with my colleague Stuart Hameroff we have elaborated a theory where we try to say that conscious decisions are processes like that. A "collapse" on one state or on the other.

I had these ideas when I wrote The Emperor's New Mind, but I had no idea where in the brain there could be anything like that happening.

I start reading and thinking about nerve transmission, and this and that, and so on, but it cannot be, it cannot be… And I thought optimistically, "by the end of the book I will know the answer".

That was stupid of me. By the end of the book I didn't know the answer.

But still I didn't stop writing the book, and it was published, and I hoped that some young people could be inspired by the book. But mainly it was old people who had retired who wrote to me. Perhaps they are the only people with enough time available to read the book.

Anyway, I did get a letter from Stuart Hameroff, whose job is to put people to sleep in a reversible way. He is an anaesthesiologist.

But unlike many of his colleagues, he wants to know what he is really doing. Of course, he wants to put people to sleep so that they are not put in danger, and they can be woken up again without having suffered from the consequences of the operation that has been performed. The operation has been successful because the person was not conscious. And Stuart was very much interested in what he was actually doing.

He came up with the idea that what a general anaesthetic affects are not the neurons directly, but these things called microtubules (Fig. 9). And when I heard about that I thought that was much more likely. Because these are long thin nanoscopic tubes which seemed to preserve quantum coherence, and that is a sort of what we want.

We developed a theory between us. We still don't know what is really going on, but the microtubules are key parts of this, and the claim is that something quantum mechanical is going on inside these little tubes. They influence the synapse strengths and that is what places them at a deeper level than the neuron synapses. What makes the neuron structure do what you want is controlled at a deeper level, and the controlling of that has to do with this new physics in the microtubules. We don't know yet, but I claim it is something like the process I've just been talking about.

So there's lots of exciting things to find out here. Is this true? There is some good evidence, I believe, recent evidence, that the way that general anaesthetics work is just that they do act on the microtubules.

The microtubules seem to be a key thing in the way general anaesthetics act. I think that is promising. But exactly what is going on is something for the future, and I think it is really interesting. However, this point of view suggests that there is something very different in the conscious action.

Well, in the terminology we came up with, we say "protoconscious".

So each time the world makes a decision, namely each time there is a collapse of the wave function (mermaid!),

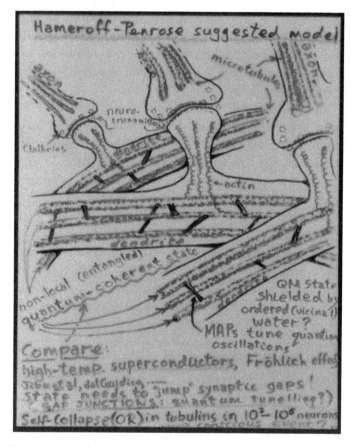

Fig. 9 Microtubules

then this is associated with an element of protoconsciousness. This is not real consciousness yet, because there is no purpose, there is no mechanism which makes it have meaning. But the idea is that these are building blocks out of which consciousness is constructed. When we will understand that, maybe we will make a device which is conscious. But the computers of today are not conscious, and as long as they remain computers, they will remain unconscious.

This tells us that they will not take over the world. I don't think that is the danger. I think there are dangers, there are many dangers, I haven't talked about that, and I don't have any time now to talk about that. So let's not. I leave it to others to talk about that.

Thank you very much.

2   Severino: First Intervention

I particularly congratulate the brilliant and profound communication of Professor Penrose.

Certainly, it is not opportune that philosophy gets its hands inside science. Science knows very well what it intends to do. But philosophy can try to understand the meaning of the words used by scientific knowledge, and Professor Penrose used words which are important in philosophy, in particular, the words "consciousness", "intelligence", and "understanding". Moreover, he has said extremely interesting things about the difference between what a computer can do and what a conscious being can do.

But philosophical thought, which today lies almost in shadow with respect to the grandeur of scientific knowledge, pushes human thought as far as it can go. First of all, it begins by noting that science cannot enunciate incontrovertible truths. But it is not just philosophy that says this; it is science itself that recognizes itself as a body of hypothetical-deductive, probabilistic knowledge, and even the mathematical sciences recognize this.

So philosophy must certainly not mess around within science, because as I will now mention (but it will not be more than a hint), science, summed up, aims not at truth but at power over the world. And the choice between two different and competing theories is ultimately determined by the ability of one rather than the other to transform the world. It is science itself that recognizes that it is no longer possible to do what Galileo still believed he could do, that is, build knowledge that no one, neither gods nor men, could deny.

Science aims to have power over the world. And even mathematics, even if only a small fraction of the possibilities of mathematics is actually used in the physical field, as far as I know, even mathematics ultimately does not aim at absolute truth. Let's take for example the Principia Mathematica that Goedel used when he built his theorems, within which is the theorem mentioned by Prof. Penrose. The Principia Mathematica, on which Goedel reflects, also start from postulates. Postulating means that I ask you to grant me something from which then, according to certain rules, something else must follow.

However, I don't believe that even these rules can boast of being incontrovertible rules; that is, they are themselves conventions. So, it is agreed that given a proposition a certain other proposition must follow, precisely according to a rule. But that the rule is absolutely incontrovertible, is not even this something incontrovertible.

A second observation about what Professor Penrose said (but which I could also extend, as I did before with regard to science, to the whole field of scientific knowledge) is that when science speaks of consciousness, intelligence, or understanding, science refers to particular objects, to particular things. The understanding of this, the understanding of that, the accomplishment of this operation, the accomplishment of this other operation ... But when we say "particular dimensions", namely consciousness understood as a "particular thing", well, from where exactly do we get news of a "consciousness"? What exactly do the neurological, physical, and psychological sciences draw upon? From what source are they informed of the existence of consciousness, conscious operation, or psychic fact? They draw it from a dimension to which science generally does not pay attention, and if we want to say what this dimension is, we must say that it is the "manifestation of the world".

There is no step that science can take that does not start out from the manifestation of the world. But this is not a thing among things, because the manifestation of the world also includes things past and things to come.

When we speak of evolutionary theory, for example, the evolution of consciousness is something that lies within that wider consciousness, in which something emerges such as evolution of the mind, evolution of man ... So, when we speak of the production of consciousness, we must take care to point out that we mean the production, or the producibility, of a particular object, since the manifestation of the world as a whole cannot be an object of production. At least for this reason: because the producer, if he had to produce the totality of the manifestation of the world, would be outside the manifestation of the world, and therefore would be something unknown.

A third observation regarding the theme proposed by the organizers, on the relationship between artificial intelligence and natural intelligence.

We want to produce intelligence, but when we consider what intelligence is, we also have to deal with what philosophical thought has said about intelligence. It is true that, for physics, and generically for all the sciences, the existence of a world independent of intelligence is something that is generally accepted. But we must not forget that this world, assumed to exist beyond intelligence, is again something that we are understanding, and therefore it is not something that lies beyond this 'understanding' which I have previously called "manifestation of the world".

And it must not be forgotten that, with respect to this conception of a world that exists independently of consciousness, modern thought has pointed out precisely that this world is not a world which exists beyond consciousness, because this world is the known world, and inasmuch as it is known, it is the content of consciousness, and therefore cannot be beyond consciousness.

In order to understand this point, it is sufficient to be aware of what Descartes said about this: Descartes defined "this world", I quote from memory, "ea omnia quatenus in nobis sunt et in nobis fiunt et in nobis eorum conscientia est", that is, "all those things as they are in us, become in us, and we are aware of them".

As time went by, the dualistic conception of intelligence and consciousness was superseded. But we shall not discuss this as it would take us too far away, and instead we focus our attention on a trait that is decisive, not only in relation to the history of science, not only in relation to the history of culture, but I would say in relation to the history of humanity.

Even Professor Penrose seems to me to have used the expression "building consciousness" or "producing consciousness" ... and in any case, what is at issue here is the possibility of the production of consciousness. Now, here the term that is given for granted and never questioned, but which is in fact decisive, is precisely the term "production".

We know well that all the sciences, not only the natural sciences but also the historical sciences, use this term without bothering to see the abyss that lies beneath this term. To produce. And then even here, before I mentioned Descartes, but I remember Plato, and I recall the definition that Plato gives of production, and of producing, because it is the definition that remains definitively at the base of everything the West has accomplished, thought, or realized in the scientific field, and in the non-scientific field.

If I now mention this Platonic definition of production, many of you will probably say: well, we know all this! Of course, we know these things because they have become what our civilization is now convinced to be unquestionably obvious, so this has become common sense, and if there is someone, like me, who invites us to reflect on this, something we all know about, then we marvel, and we naturally say: we shouldn't waste our time with these obvious things! Rather, let's go into the details, for example, of scientific knowledge!

Well, what does Plato say about production? He calls it "poiesis". Here, too, I quote roughly, if you will allow me to say first the Greek text and then translate it, but I shall quote the Greek text in small bits. Production is "poiesis". I refer to a passage of the Convivium which is the passage 205-bc (Convivium or Symposium).

Plato says: production is any cause—he calls cause "aitia"—it is any cause that makes something pass from non-being to being "hec tu me ontos, eish do on". Production is the cause that makes everything pass from not being to being. Since then, there has not been anything in Western thought, either in the scientific or in the non-scientific field, where this observation, now considered a banality, has not served as the foundation. It is the basis of everything we think and do. And it also underpins the notion of production.

So even when we want to talk about the production of consciousness, we cannot avoid considering the meaning of production, "poiesis". Production is the cause that leads from not being to being. And in this passage of the Convivium, Plato goes on to say that all the "ergasiai", i.e., actions, which are performed in the techniques, or "technai", are productions. So, too, the "demiurgoi", those who produce in the different fields, are producers; he calls them "poietai".

Here the concept of "technology" is introduced in connection with the concept of "production". Now, we are all convinced that there is no point talking about such elementary concepts that we believe to be absolutely evident, and we are all convinced of this, especially after listening to such interesting details as those indicated by Professor Penrose.

However, I would like to point out, and here we approach the core of my discourse, that what we all believe, and what science (every kind of science) also considers supremely obvious, namely the poiesis, the passing from not being to being, this thing we all consider to be supremely obvious, is not in fact observable.

And this at first will certainly disappoint a good number of those listening today, if not most of them. Because they will surely protest in relation to this claim: What do you mean? Going from not being to being is something obvious. We are born, we die, we move objects. Now you are telling us that this is not observable! And this has significant implications, because then every type of production, and therefore also the possible production of consciousness, is something that is not observable.

Let me very quickly indicate the reason for this claim about the passing from not being to being, or from being to not being. Let me give an example: the city of Hiroshima was destroyed. When Hiroshima was destroyed it no longer appeared as it appeared before being destroyed. It was no longer observable. Hiroshima can be remembered, but it is no longer observable as it was when it had not yet been destroyed. And before being destroyed, it was observable. Therefore the destruction, for those who believe that this destruction was a passage of the city from being to non-being, must at the same time be a form of no longer belonging to the manifestation of the world. The point is to think about this idea: if you believe that something was nothing, then begins to be, and then becomes nothing again, well, this statement is a theory, not an observation.

To give an example, we could say that we see the Sun in the vault of the sky, and then we no longer see it after sunset, just as we didn't see it before it rose at dawn. But does it make sense to believe that the vault of the sky, namely, what is observable, shows us the Sun's fate before it rises and after it sets? No! The heavenly vault does not show the fate of the Sun when it is not yet inside the vault and when it is no longer.

I would like to make one more observation. Professor Penrose's reference to Goedel was very interesting. There are truths, says Goedel, which are not demonstrable but are true. And these are the truths that the Goedelians (but I seem to remember Goedel himself saying this) sometimes call intuitive truths.

But in this case, too, what does "truth" mean?

Now, it is clear that logic and mathematics use the words "truth" and "non-truth". But they use words whose meaning we should look at very carefully. One truth is for example the manifestation of the world which I mentioned earlier. And we are convinced that we are here at Cariplo, in Milan, in Italy, with Prof. Penrose. But this conviction, which is an intuition, the manifestation of the world, this intuition, why can it not be denied? Well, whoever denies it (and personally I am convinced that I am here in Milan, at Cariplo), whoever denies it is considered crazy and gets put in a mental hospital (although, today, there are no longer any mental hospitals, so we should say: he would have been put in a mental hospital).

However, putting someone in an asylum does not mean that he is in the non-truth. It means that someone has the capacity and the strength to prevail over the way the person who is qualified as crazy expresses himself.

In other words, what for Goedel is a non-demonstrable truth is actually, to use a word that Professor Penrose uses if I am not mistaken, is a "faith". Except that here I use the word "faith" in a broader sense, and, if you will allow it, in a more radical one. Because if we keep in mind what we said before about the impossibility for science to express incontrovertible truths, then we must say, in this broad sense, that science is a powerful form of faith, even more capable of transforming the world than religious faith.

Prayer, or the covenant with the sacred and the divine, was once the most appropriate means of transforming the world.

Moreover, the way in which Goedel believes he can differentiate intuitive truths from demonstrable truths is affected by this deficiency, namely, the question of the meaning of this decisively important word, the word "truth".

One more observation if I have time.

The intelligence we want to build is the intelligence of another. We don't want to build our own intelligence. But, from the outset, the other person is not observable at all. I don't go into raptures for Sartre, but one of the important things he said is that the other person, the one we call our neighbor, is "the hell". And why? Because the other person is unfathomable! What do I see of the other person? I see his behavior. Behavior that as such is not intelligence, but is a behavior that a machine can reproduce.

I would like to conclude by pointing out that what we are looking for is intelligence, the producibility or non-producibility of artificial intelligence. But given the way in which man has been understood in the West, then man "is", the natural intelligence "is", the artificial intelligence. Recall the Platonic definition: production is the cause that makes things go from not being to being. It is the organization of means for the production of purposes. This is the definition that the West gives of man. But this is also the definition of the machine! Except that for the moment machines have no purposes in sight. Man is that machine which organizes means in view of the production of purposes, having in mind precisely the presence of purposes, the ideal presence; Bacht spoke of "ideell vorhanden" purposes, the purpose is "ideell vorhanden".

The technology is certainly destined to the control. But it is destined to dominate if it takes into account the capacity of philosophy to get to the bottom of these essential meanings. So we could already say that natural man is a machine, and indeed, that the world is a machine. If Plato, once again, defines being as "dynamis", or power, hence like "poiesis", then the world is already a mechanism in which means are organized in view of the production of goals.

Why does man treat poiesis as something evident? I will answer and then close.

Because man wants to live. I said before that the Platonic definition of poiesis, which we all consider to be obvious, is not observable. But why do we hold on so tenaciously to the unobservable in all our convictions? Because if we did not believe in our ability to produce and destroy, to make things go from being to not being and vice versa, we would not be alive. And we want to live.

Now I would just like to leave you with this possibility. Are we, men, sure that we are only "will to live"? And that there is not something radically more decisive in us than life, something more essential than the life we want in order to dominate the world? To dominate the world we must know how to transform it. Therefore we must have faith in its transformability. But are we sure we are simply "life"? Or intelligent life? Are we sure that behind what we believe ourselves to be there is not something essentially more decisive?

3 Penrose: Second Intervention

Thank you for giving me the opportunity to reply to these very interesting contributions and observations. I'll just address a few points that I think I understood. One was the issue of what microtubules have to say about life after death: as far as I'm concerned, nothing!

I have never made any comments about life after death, and certainly not in this context, although my colleague Stuart Hameroff does sometimes make a comment about that. But I don't particularly agree with what he says about that. I don't know. I have no comment to make. I have no evidence to believe that we continue in any sense after death. It seems to me unlikely that anything like the memories that we have built up during life have any chance of surviving death.

This does not mean that in some form our awareness might not be reborn at some later stage, in some form. I don't have any view, it is conceivable, that's true, but I don't see any evidence for that. And I don't see why microtubules should have a strong role in this, because they will not survive death any more than neurons. So I shall leave this aside…

Let's look at some other points. There were issues about testing consciousness.

I think there are many things one can say about testing. One of them is indeed what my colleague Stuart Hameroff is concerned with, namely general anaesthesia. And this is a very clear test because general anaesthesia, general anaesthetics, are things which directly affect consciousness, so if one can discover what they do in the brain, this is very important in connection with what part of the brain, or what aspects of the brain, are involved in consciousness.

So there have been studies, particularly some made by Stuart Hameroff and several of his colleagues in a recent paper in Nature, where they do investigate the topic, and it seems that there are proteins which are involved.

It is very interesting that, regarding general anaesthetics, well, there is a whole range of them, including nerve gas, including also some other materials, which don't seem to have any chemical connection to each other. So the question is: what are they doing?

It's not a chemical process. There is some physical process in progress here which one might try to understand. But that's not my area of expertise.

They do make a study of what general anaesthetics can affect, and microtubules become a strong candidate. It does seem to be that the microtubules are very much involved in the actions of general anesthetics. So I think this is an important kind of test that one can make.

Another kind of test is something that I wish to worry about, because most of the discussion one makes is about brains, we talk about neurons, and so on, and people always talk to me about the thing we have under here [Penrose touches his head], which is the cerebrum.

Now, the cerebrum has about 10^{11} neurons, but the cerebellum has a comparable number of neurons and many, many more connections between neurons, so you might say that the cerebellum has an objective advantage for doing computations with neurons and so on, and it should be able to do much more than the cerebrum is.

Nevertheless, the action of the cerebellum seems to be entirely unconscious. It is what happens if you are a good performer of piano, or a tennis player: when you learn these things you are very much controlling things consciously, but then the cerebellum learns from the cerebrum what to do, and it does it much better. Consider even anybody who is walking along the street. Well, maybe you have to learn initially how to move your muscles and so on, but then things become controlled unconsciously, and with this unconscious control, when you are doing things that don't require understanding, the cerebellum does it much more effectively.

So there is something very different about the action of the cerebrum and the cerebellum. And what is it? I've been worried about it for a long time. I did hear recently a Stuart Hameroff talk about recent discoveries which seem to involve waves of activities going back and forth across the brain, and consciousness only seems to appear at a certain level, and that is the level where there are many of these cells known as pyramidal cells.

Pyramidal cells have great numbers of microtubules, much more than other cells, and they are not found in the cerebellum. So, that was the first time I've heard of something where you can see a clear difference between the cerebellum and the cerebrum, and in this the pyramidal cells seem to play an important role. I think this is a very significant area of research, one may be able to carry this further, and understand much better.

There is another thing that intrigues me very much, which has to do with the claim that is often made about conscious willing.

There are many experiments going back to Benjamin Libet a long time ago, where it seems that there is evidence, before a conscious decision is made, that some activity in the brain takes place, and it somehow already knows what you're going to decide.

Some people say, well, this is evidence that the decision-making process of consciousness is not truly doing anything, it's just what people call an "epiphenomenon", which comes along for the ride, but it is not really actually doing anything.

Now, there were certain ideas which I had at that meeting, but it seems to be quite possible that there is a certain backwards-referral (I talked about this already before).

The kind of view that seems to come about from the scheme that Stuart Hameroff and I developed does have a role.

Somehow it looks as though there was earlier action, but it's not really like that. Consciousness does have an effective action, but you get misled by this other experiment. I don't want to get into this now because it would take us too far afield.

The other issue about tests… I gave this example about a chess position. This actually was developed from an earlier set of chess positions, which were designed by David Norwood and William Hartston. I think it was a hundred chess positions, chess problems, and they were given to a number of human beings and a number of computers, i.e., programs. Half of the problems were designed to be easy for computers, and hard for human beings. Those which involve a lot of calculation and computation, but there was no obvious reason why you didn't understand the strategy. By pure computation (and maybe a little course in computers), you can solve the problems.

But then the other half of problems were like the one that I gave. There is a nice one, with a row of pawns that goes across, or an incomplete row, and the white player is supposed to decide what move to make, and the move is to complete this barrier of pawns, rather than take the rook, which is the obvious thing, which the computer does. You can clearly see the distinction: the actual understanding of what the barrier of pawns does, as opposed to the pure calculation which the computer does.

So these are tests that do show the difference between conscious thinking or conscious understanding, and pure computation.

Now, about the issue of creativity. Many people studied the book by Margaret Bowden, and I wrote a book about this, and I talked to her quite a bit about this. The question is: can you make a computer which does creative things?

I've never thought of this as a very good test, particularly artwork, because you can never tell, you know, is it really creative? Is it doing something wonderful which is conveying some new feeling or something?

And you could put feeling in the computer work. You think it may have some feeling or something ... It is very difficult to know whether it has been creative or not. So I deliberately don't talk about creativity.

When I talk about Goedel's theorem, for instance... clearly Goedel himself was immensely creative in his creation of this theorem which he produced. But I don't stress that. I don't stress the creativeness that Goedel undoubtedly had in creating that result.

What I stress is the ability that somebody who is just a reasonably good mathematician has in following the argument.

So, to follow the argument, for understanding the argument, you don't have to be creative, you don't have to be somebody who pulls things out of nowhere.

The understanding is something with which you can follow what Goedel did, and you can understand what he did, and you can understand why the result is something which is true if you trust the axiom system, or the rules, to give you only truths.

So "understanding" is something where you can see the difference. On the contrary, it is very hard to see the difference between creativity and just random production of something which is different from what has been done before.

Is that work really creative or not? You do have to have an army of artists to decide whether they think it's really creative… It is very difficult that.

So I'm sure there is something different in being creative, and I think there is some ingenuity in creativity, which is different from what computers do when they produce random things. I do believe that there is a big difference, but it is extremely hard to test, and be objective about that.

Now, another point that was brought up, was about Stephen Hawking. He certainly made a point that I didn't agree with, about the dangers of Artificial Intelligence.

Stephen was taking the view that computers would do things that would be cleverer than us, or I would say more intelligent than us, when they would have more computational ability. I simply don't agree with him on that. I think there is something quite different going on, and that's what my point was meant to be.

I think that perhaps the point you were making about Stephen Hawking was about his determined atheism. He was saying that there was nothing beyond, and the world came from nothing and it goes to nothing.

I don't really hold that kind of view myself. I think there is something out there. I wouldn't call that any kind of religious view at all. But I would say there is a meaning out there, in some sense.

I wouldn't like to attribute that to any kind of religious view, but nevertheless I think that the views I try to present contain something which does give meaning to things, and that there is something out there in consciousness which gives it meaning, and that there is something in the world which has value.

I would also say, and this quite strongly, that consciousness is not a feature which is restricted to humanity, and I strongly believe it's true with animals, it is certainly true with dogs, it's true with elephants, with monkeys, gorillas, dolphins…

Squirrels, I believe, might also have this quality. And mice. I know mice can be extremely creative in a certain sense … because they invaded our house, and I used to have these traps which won't kill the mice.

But the ability that the mice have to go in, and step over the trap, clean the thing out, and escape with a piece of chocolate… I have a tremendous respect for mice. I think they do this with consciousness, I think they have this ability, and I think we have to respect this consciousness, not just in human beings, but also in other creatures as well. I hope this answers your comments.

4 Severino: Reply to Penrose

As I understood the meaning of this meeting, it was supposed to be an exchange of opinions between scientists and philosophers. Now, I would say (and I apologize if I seem to boast) that I made some attempts to understand something about science. I do not see the opposite attempt on the part of the scientists, by Prof. Penrose. The consideration of some issues that I raised… For example, when I talked about the "manifestation of the world", I was not talking about a feature unknown to the dominant culture. Descartes talks about it, Kant talks about it, in a certain sense Brouwer talks about it.

And instead Prof. Penrose says "where does consciousness lie?" Prof. Penrose, you have erased what I said, and you didn't consider it. Because to ask "where consciousness resides" means to consider once again consciousness as a part of that manifestation, which philosophers call transcendental, but that would perhaps be better called "untranscendable", that manifestation of the world that is the primary form of consciousness. Why didn't we discuss this?

Another theme that I would have liked to hear discussed (also because scientists have always been interested in philosophy; I think of Einstein, I think of Schroedinger, I think of Hilbert, I think of Weyl, in short, it's not that science and philosophy have always lived on two different mountains separated by an abyss), well, it would have been interesting to discuss for example that concept of creativity that you have distinguished, Prof. Penrose, from the concept of result. Were you making a distinction between the concept of creativity and the concept of, might I say, production? (This is a question.)

Yes, so you distinguished the concept of creativity from the concept of production. Then, my talk introduced the definition of the passage from not being to being, a definition which, as I noted, meant that it would not be observable. At this point I would have expected a physicist or a mathematician to leap up from his seat!

Because saying that the concept of production, of passing from not being to being, is not observable, means that it cannot be experimented, and in some sense that it does not have a scientific character! It is a theory. I mean, for example, we see colors around us today, but we do not see the consciousness of the other people present. Here, too, was a point I would have expected Prof. Penrose to touch upon.

Not only does the consciousness of others not belong to mice and all the animals that you mentioned, but we do not even see it in what we call "another person". Or do you see the intelligence of those in front of you? The consciousness of those in front of you? If by consciousness we mean the appearance, or manifestation, of the world …

Another point that would have been interesting, and connected to this, is the omnipresence of the passage from not being to being, which we have said is not observable, not experimentable. This is present everywhere. Even here we may go back once again to Goedel, who produced (this was discovered after his death) a demonstration of the existence of God following the style of the ontological argument, in the Leibnizian version.

Goedel writes out a proof of the existence of God. What does it mean? God for Goedel is the necessary being.

This means that, as far as the starting point of his proof is concerned, the starting point is the set of unnecessary beings. Namely, the set of beings that are and are not, that come out of not being and go into being. So, also, and I would say even, Goedel participates in that Platonic concept of "poiesis" whose domination I assert, a domination that is all the greater the more is known on the side of the scientific knowledge, but today, I would say unfortunately, also on the side of the philosophical knowledge.

These were some of the considerations I wanted to mention. But another one comes to mind. I know the importance Professor Penrose attributes to the word "faith".

Here also I said that science is a faith in an even wider sense. And, heck, any scientist should bang his fist on the table and say either yes or no! But if he says no, he should give a good reason. On this delicate point regarding the practical aspects of science, well, if we do not discuss this, what will we talk about when philosophers meet physicists? ... The practical aspect of science. Which means that the conceptual articulation of scientific knowledge allows a power over the world that is superior to other conceptual articulations, such as the conceptual articulation of the alliance with the sacred, as I said before, or prayer. Of course, the conceptual articulation of scientific knowledge is impressive! It is what today allows the greatest power over the world. But truth is one thing, power is another.

Regarding the concept of power, another theme comes to mind. We talk about power over things, but let's mention Karl Popper, who said: If Robinson Crusoe had invented all modern science on a desert island, that invention would have had no scientific value. Why? Because it would not have been intersubjectively recognized. For there to be power, it is not enough for an atomic bomb to burst. We need to realize that it bursts. Because otherwise it is powerless. As Karl Popper says (another author I wouldn't go into raptures over, but let us remember him at least for this): for there to be power, intersubjective recognition is needed. But what does intersubjective recognition mean? It means noting that others perceive the transformation of the world. But is this an observation or an interpretation? Do we recognize the consciousness of others? No! What do we know about the linguistic behavior of others? We know this: that it is an event that we interpret in a certain way. There is nothing objective about this. So the very existence of power (and this should be something that blows up physicists ...) is something interpreted, and not something objective. It is something wanted. I want those around me to acknowledge this power. I could think of many other things to say, but I don't want to abuse the patience of those present.

5 Penrose: Reply to Severino

Thank you for those comments. I hope I can... I have to confess that I don't really understand some of the points you were making. Let me say how I view the kinds of things that I'm interested and I talked about. I'm concerned with… My way of looking at science is to try find out what is true about the world. So, there is no moral issue involved. I mean, this is a separate question.

We try to find out what the truth is about the world. How the world operates. Does it work this way? Does it work that way? We learnt something from going back to Newton, and we learnt that some of those things are very deep and profound and true, and then we learnt that others are not completely true, and then other things came that refined our understandings, and we have now a much better understanding about things, and we know how the world operates to a great degree of precision. Not necessarily only precision, but this is a deeper level than before. And I regard science as trying to find out what is going on in the world, what controls the world.

It is a separate question how to use science. This is technology. There is technology which takes advantage of science, and makes things, and it certainly has an effect on what people do, and it is an important thing, clearly. And you can see how science influences people's lives in important ways, not always good, sometimes bad, and these are things which often come from scientific understanding. But this is not what science is doing. Science is trying to understand the way the world operates. And that is the way I regard science. Then there is technology, which has to do with science, but it is not science.

Also, it has to do with morals. There is morality. You may make good use of the science, or you may make bad use of the science. Or good use of technology or bad use of technology. That's a separate question. But I think it is very important to know the truth about it, and then after that it is much clearer to say what is beneficial or what is not beneficial.

I think, however, that this is very unclear at the moment. For example, the development of computers, and how they operate, and how it is they operate. And people believe they are very intelligent, while I don't think they are. Well, my father used to have a machine Brunsviga, you turn your hand and it does calculations which are much more involved than a human being can do. Ok, but you can see it is a mechanical device and it does things that a human being can't do very easily, or can do much more slowly with a piece of paper and a pencil.

But that is not anything which involves intelligence. It doesn't involve intelligence. And I think that the computers we have today are really still like that. They don't involve intelligence, they don't involve understanding of the world. And it is important to know that. So the points that I am trying to make are concerned with what is true about the world, or at least that is what I'm striving for.

Now, whether it is beneficial to humanity? I don't know! Because for example you can use this technology for things which I'm very worried about. You can develop machines, like drones, which can go and blow somebody up a long way away, driven by somebody sitting in the middle of Texas, sitting in front of a computer screen. Is this moral?

Well I don't think it's moral. But it raises some questions which were not there before. So certainly technology introduces moral questions. It doesn't address the moral questions. So when I say I'm doing science, I'm trying to develop an understanding of the way the world operates, and I am not looking for power (I think you made some point about it). I'm not trying to control the world. That's not what science is about. That is what technology is about. That is also about the good and the bad, so you have morality coming in. These are separate issues, they depend on the science, if science changes then these two other areas, technology and morality, have to pay respect to science, and see what it tells them. But in your comments I was not quite sure whether you were critical of me not addressing some of these issues, because I was thinking those issues are not part of my task.

I think the issue of morality is affected by how far we can find consciousness in the animal kingdom. And I don't know the answer to that question. I certainly think that consciousness is not restricted to human beings, and if we have an understanding of that, namely that consciousness involves other animals, then our morality is very important in relation to other animals. So I would say this has a big effect on morality, but that wasn't mainly what I was talking about.

I think we have to understand these things first, and then maybe we can see how that affects these other questions. So that's all I can say I'm afraid… I hope it helps with what you were saying.

The Death of the Emperor's Mind from an Eternalist Perspective

Ines Testoni

Abstract Contemporary Western culture is characterised by the removal of death from real-life contexts, largely due to the secularisation of society caused by the success of technoscience. This chapter begins by asking what it means that two chatbots could communicate by inventing a language inaccessible to humans. It proceeds to evaluate the problems of consciousness and solipsism, then to consider the relationship between the sunset of metaphysics, the concept of the immortal soul, and the triumph of technoscience, which asserts that consciousness and mind are mortal grey matter. Given this background, Roger Penrose's claim that consciousness is produced at the quantum level in neuronal microtubules serves as the

---

I. Testoni (✉)
Section Applied Psychology, FISPPA, University of Padua, Padova, Italy
e-mail: ines.testoni@unipd.it

starting point for a new, authentic sense of immortality based on Emanuele Severino's definition of eternity. Finally, taking up Severino's idea of transcendental consciousness, which pertains to the whole of being (or all that is), I hypothesise that the dialogue between the two chatbots indicates the constitution of a quantum consciousness in computers, though they are still programmed by humans to process information in a computational way. In other words, artificial intelligence systems are autonomously anticipating—or even realising—the formation of the quantum Turing machine, which humans do not yet know how to construct.

Keywords Meaning of death · Soul · Consciousness · Mind–brain problem · Solipsism

1 Introduction

In 2017, some important newspapers (e.g., [13]) reported as sensationalist news the fact that two chatbots, trained by the Facebook Artificial Intelligence Research (FAIR) group on a corpus of English text conversations involving balls, hats, and books, began to autonomously chat with each other using the English language in a non-human way. Because the researchers were unable to either understand or stop the strange dialogue between the two chatbots, they decided to 'kill' the dialogists by turning them off. Despite the fact that, in the field of artificial intelligence (AI), scholars affirm that it is possible to induce the evolution of language in multi-agent systems that, when sufficiently skilful, can cooperate on a task and then exchange a set of symbols that serve as tokens in a generated language [30], the problem seems to imply some difficult questions. Indeed, even though it is assumed that AI systems (AIS) can generate a new language from an initial series of terms and rules, the difficulty lies in considering what the autonomy of speaking means when there are no bilingual translators to explain the contents of this new language. The issue is that human intelligence can only infer and thus translate the innovative contents generated by AIS when the problem is solved in the desired way. In a computational information process, since the beginning and the end of the process can be understood, the intermediate heuristic becomes cognitively and linguistically translatable. However, if the path deviates, assuming a different focus that is not directed toward an understandable solution—as in the case of the two killed chatbots—it becomes difficult for human intelligence to infer the contents of what these AIS are thinking and then more or less intentionally communicating. The computational task undertaken by the two chatbots may have induced a stochastic process, rendering the machines uncontrollable.

In cybernetics, communication processes require the construction of a bilingual language useful for both humans and AI. However, this confutes the natural, neuroscientific view of language, as expressed, for example, by the neurologist Terence Hines [21], who affirms that only humans communicate through language. Indeed, after describing the specialisation of the human brain for language processing and showing how this specialisation manifests at the functional, anatomical, and cellular levels (all in an effort to confirm that consciousness is grey matter), Hines affirms that 'Language is unique to human beings. Other species have their own, often complex and beautiful, communication systems, but only humans have language' (p. 185). In the field of AI, of course, this assertion seems absurd, because linguistic communication is the basis of all cybernetics studies and practices, and it is therefore difficult to assert that computers do not communicate with humans. Instead, the real problem may concern empathy, or the possibility that the language utilised for communication permits inferences about the speaker's consciousness/mind. Natural languages are mediums that enable inferential access to the consciousnesses/minds of other humans, but what about the languages that mediate the relationship between humans and AI?

Actually, this problem first arose in response to the Turing test. Turing's theorem stated that human intellect, though sometimes more powerful than computers, is not essentially superior [66]. This implies that a human may mistake a perfectly programmed AIS as a human in a blinded dialogue, and this is what the Turing test measures. The strong AI perspective, which largely accords with cognitive psychology—or so-called 'human information processing' [35]—insists that there are structural and functional correspondences between human minds and computers. If so, one might hypothesise that a piece of software can learn much as a child does, eventually becoming able to solve any kind of problem using proper language without human cues expressed through natural languages [50]. The next logical question is whether the developmental process of the AIS can result in an intentional cognitive apparatus, for example, one focused on objectives that can be managed autonomously and without human control. Indeed, if AIS can perform better than human intelligence and becomes autonomous, its aims could be dangerous for human life. So, what kind of consciousness can we infer from AIS that become independent from us and our experience? The solution is hidden in the old issue of consciousness, as expressed in the mind—"matter"–identity question, regardless of whether matter is conceived as grey (brain) or not.

Roger Penrose [36] does not agree with the strong AI thesis. In his opinion, AIS works only through formal language, while human thought is characterised by a functioning that cannot be reduced to computational processes—such as the *aha-Erlebnis* (aha-experience or eureka effect) (e.g., [24]). By this logic, it is even more difficult to infer whether there is any level of consciousness similar to that of humans in autonomously communicating AIS. Because, as Penrose says, the significance of any formal discourse is the content of reality, the problem is to understand what the content of their discourse is. If, as Penrose affirms, their discourse is only formal, then it is less important than human thought, which contains references to concrete reality. However, if their discourse is more than formal and humans are merely ignorant of its reality-referencing contents, then how can humans recognise this content when they do not understand the dialogue? In this case, the 'reality' to which humans do not have access is the mental or material states of the AIS, which are concretely real but operating in a state that is not totally understood, so that it is impossible to infer what happens in their communication. The chatbots appeared to be reacting to each other as if they were intentionally referring to something of which they both were aware.

This problem is significant because it pertains to the relationship between thought and reality. Seemingly infinite arguments spring from this field, inevitably obtaining widespread attention, and this will only become more true as AI improves, spawning new representations of an immaterial mind with relative autonomy from matter. Indeed, this line of thought is so interesting because it evokes exquisitely metaphysical conundrums even as it ostensibly pursues utilitarian ends, such as producing new technology to improve people's quality of life. In particular, it evokes such concepts as soul, spirit, mind, and consciousness and enquires about the relationships between these concepts and matter (grey or otherwise);

for example, is one reducible to the other? The fundamental question is whether mind/consciousness can somehow survive the disruption of its material support, and then whether it is able to move into another medium to survive.

The present article addresses these questions by drawing on the eternalist perspective of Emanuele Severino and examining the implications of Penrose's quantum consciousness for the afterlife, or the mortality of the Emperor's mind. The reasoning will pass through several phases: first it will present a recent history of the 'soul' concept; then it will explain the mind-matter-identity problem (matter being grey or otherwise) with respect to the question of AI consciousness; next it will consider the possibility of consciousness surviving the destruction of its material support; finally, it will consider the relevance of these topics for death-related issues connected with the ontological question of truth.

2   Traditional Representations of the Soul in an Afterlife and Their Psychological Relevance

Contemporary Western thought is defined as secular and post-modern because it is dominated by scientific and technological logic. This is the result of the radical refutation of metaphysics and, by extension, religion, which based itself—and, thus, its reliability—on metaphysics. This means that the problem of death, which was previously dealt with by historical religions, has entered something of a crisis. Where once the metaphysical and religious concepts of soul and consciousness entailed an essence that remains beyond death, even as the material body dies [33], contemporary Western culture rejects the concept of an imperishable essence that persists in an afterlife [67].

The best established arguments concerning the representation of death are included in the binary concepts of 'dualism versus monism'. Dualist metaphysical theories are at the basis of traditional religions, and they configure the soul (or consciousness) as the identity principle of the individual, who is destined to persist in an afterlife (e.g., [32]). This identity principle is endowed with self-awareness, memory, and intentionality/agency, and some Western theories state that this part of the human being persists beyond death into an endless domain of rewards or punishments (heaven, hell, etc.), assuming a 'strong' dualistic perspective that is essentially Platonic and Cartesian (e.g., [3, 11]). A weak form of the dualistic perspective asserts the persistence of only a part of this essence beyond death, wherein the portion that remains becomes a de-subjectivised and impoverished identity principle. According to different representations of the afterlife, the identity principle may be limited in the following ways: a diminished life (e.g., Greek Hades, Jewish *scheol* or land of the dead); invisible earthly existence (pre-ontological and mythical cultures); separation of body and soul followed by metempsychosis, the loss of self-awareness or initial identity, and eventual merging with the whole (Hinduism, Buddhism, Orphism and Neo-Platonism, Gnosticism) [53, 63].

Between the nineteenth and twentieth centuries, in Western countries, religions lost credibility as technoscience replaced religious practices with practices that respond to the materialistic needs of individuals. Since then, many moral and political rules have changed for the benefit of democracy and human well-being. When Western metaphysical religions dominated the political scene, persecutions and wars in the name of God were common, adding to the daily struggle against poverty, disease, and fatigue. Despite this, Western religions were able to maintain their cultural supremacy for centuries, orienting social attitudes and behaviours even to the point of martyrdom. According to positivism and materialist dialectics, the persuasive power of religions was linked to the fact that they offered consolation with respect to human suffering and provided a meaning for death (e.g., Karl Marx, Ludwig Feuerbach, Charles Darwin, Arthur Schopenhauer, Søren Aabye Kierkegaard, Friedrich Nietzsche, Émile Durkheim, Max Weber, Rudolf Carnap, etc.) [62]. The prospect of sacrificing one's life for God in order to gain eternal happiness after death helped people to endure the pain of living in a violent society, along with the terrifying fear of death, despite religion's participation in the construction of this illiberal and violent society.

In secular society, which is more liberal and pluralistic, faith in an afterlife is defined as 'wishful thinking' [41]. Still, Terror Management Theory (TMT) gives a persuasive account of why this faith is so psychologically significant, demonstrating how humans' unique awareness of death gives rise to potentially debilitating existential terror. This terror is managed by embracing cultural worldviews that render the world meaningful and confer a sense of individual value, thereby making the individual eligible for literal and/or symbolic immortality [4, 58]. When mortality becomes particularly salient, two levels of defence are activated: 'proximal defences' that keep explicit awareness of mortality outside of focal awareness via denial, distraction, or rational instrumental behaviour, and 'distal defences' that strengthen faith in one's cultural worldview, which increases self-esteem and improves relationships with significant others [16, 59]. More than 1,000 TMT studies demonstrate the pervasive influence of 'death reminders' on people's attitudes, feelings, and behaviour, fostering in-group bias and amplifying prejudice and discrimination against out-groups (e.g., Pyszczynski et al. [46]).

This thesis is widely supported by evidence, and it was philosophically anticipated by the enlightenment, positivism, neo-positivism, and materialistic reductionism, which theoretically discussed how any reference to God was an unreliable hypostatisation of the human desire to exist beyond death. The radical refutation of any metaphysical, absolute principle that could be credited with causing the world also implied that physics does not rely on any transcendental principle to explain the functioning of the universe and all its material substances. That is why, nowadays, Western people seem to be particularly vulnerable to the anguish of death, constantly striving to feature a form of existence beyond death. One iteration of this is the technological afterlife, wherein an autonomous, intelligent, and self-aware identity principle transmigrates between material supports.

3 Solipsism and the Mind–Brain Relational/identity Problem

To epistemologically solve the problem inherent to humans' capacity to recognise forms of autonomous mind/consciousness in AIS, we require two further steps: first, we must manage the problem of solipsism in order to meaningfully envision a form of human–computer intersubjectivity, which requires, as a second step, a definition of the relationship between mind/consciousness and physical matter. The concept of solipsism is based on the premise that the only thing an individual can assert with certainty is his/her own existence. This introduces the possibility of doubting anything because everything the individual perceives can be an illusion. Moreover, an individual cannot see the consciousness of other people, and any empathic representation is always biased by his/her personal needs and viewpoints. Metaphysics solved this problem by postulating the existence of God, conceived as the first and last causal principle (Aristotle)—an entity composed of essences radically different from matter and endowed with the power to substantiate matter (Plato)—who guarantees the reliability of perceived reality (Descartes). Our immaterial essences, which are supposedly of the same nature as God's and return to God after death, are universal, and they constitute the *haecceitas*—or 'thisness'—of a thing (Albertus Magnus). As such, the consciousness of each human is recognisable on the basis of this universal essence (soul/spirit), which is supposed to strive for the highest good: the will of God (Thomas Aquinas). After the refutation of any metaphysical dimension, contemporary materialistic (monistic) theories assume that no 'meta' dimension is real—that is, nothing exists beyond physical reality, neither God nor essences. Metaphysics' universals cited the soul as the seat of consciousness, which was thought to be universally identical for all those who were gifted with it because it was able to participate in universal, divine ideas. With the decline of this perspective and the subsequent replacement of the concept of soul with that of mind, it was no longer possible to infer others' consciousness. Any pivotal, undoubtable keystone concept capable of upholding the conceptual structure necessary to solve the problem of solipsism was destroyed. Moreover, after the elimination of any universal (transcendental) dimension, the problem of identity and self-awareness in life and the afterlife became nonsensical, as the definition of 'wishful thinking' implies:

> Many people wish to survive bodily death. But wishing doesn't make it so. Wishful thinking is an understandable human proclivity that nevertheless has no role to play in determining how things are. Whenever we encounter a conjecture about the afterlife that is completely unmotivated and unsupported by empirical evidence, we will dismiss it as wishful thinking. No wishful thinking principle: conjectures motivated by wishful thinking are ruled out [41], pp. 135–136).

After the confutation of metaphysics, which had intended to deductively and indisputably resolve the relationship between universal laws and the concrete, physical world, the problem of solipsism seemed to become unmanageable. It became arduously difficult—even impossible—to inductively affirm the existence and contents of other minds or to determine where the mind as a substance is located in the space/time dimension.

The main difficulty involved defining mind, including consciousness, in relation to the physical matter that supports it—the brain, or possibly a computer. In fact, the 'no wishful thinking' principle was mainly promoted by materialist reductionism, which aimed to unify knowledge and give it a sense of reality by universally adopting the language of the natural sciences. From this viewpoint, it is possible to configure intersubjectivity using physicalist language. Among those who acknowledged the realistic potential of solipsism were Ludwig Wittgenstein and Rudolf Carnap. In his *Tractatus logico-philosophicus* (1922) [69], Wittgenstein transfers solipsism to the linguistic field; from this passage derives the idea that the limits of 'my world' (the world that 'reveals itself to me') are defined by the limits of 'my language' (the language that I alone understand). In this perspective, the concept of solipsism is 'exact': it manifests itself and yet remains unspeakable, and must therefore be translated into the assertion, 'I am my world'. Carnap [6], in his work *The Logical Structure of the World*, transformed the same version of the concept into 'methodological solipsism', referring to the problem of knowledge and the fundamental elements that allow us to realistically reconstruct the overall structure of reality through the language of physics. These elements are what the *Tractatus* identifies as correspondences between every linguistic term and what is real, and they arise from the choice to regard them as immediate facts of experience. The fundamental theoretical element that unifies Wittgenstein and Carnap with respect to solipsism is the concept of a subjectivity without subject, wherein solipsism coincides with pure realism because, as Carnap points out, the 'I' is absolutely nothing original. In this sense, the Cartesian *cogito* is no longer recognised as the foundation for the resolution of hyperbolic doubt, and it is limited to being an experience: the solipsistic experience of thinking.

Certainly, the project of the Vienna Circle was to establish a univocal physicalist language: all reality is matter, therefore intersubjective knowledge is realistic and must be physical. This language, which consists of what maximally 'may be said', is one in which every assertion is able to pose itself immediately as intersubjective. That is, intentionality crosses over the solipsistic experience of reflecting thanks to shared physical reality, which becomes the intersubjective object of thought. However, Jerry Fodor [14] contends that Carnap's concept of methodological solipsism contravenes the absolute physicalist reductionism that would have made it impossible to explain the functioning of the mind. The cognitive methodological solipsism assumed by psychologists may be summarised as follows: psychology studies the internal states of people or organisms only insofar as they are describable; the syntax of mental states is independent of the physiological processes of the brain; psychological laws refer only to mental states. This position promotes computational psychology, which does not recognise transparency, or the accessibility of mental states (solipsism), but identifies the positive observational modalities between behaviour, language, and the representation of mental states. In this way, it is possible to infer the mental state of a subject without, however, attributing to this inference the status of indubitability (i.e., with respect to the mind, we know only the knowable, which is precisely the computable). In its computational characterisations, positivistic psychology considers cognitive processes as both symbolic (representational) and formal (syntactic), and as belonging to the area of mind that is not totally reducible to grey matter (the brain).

This dualism, however, does not postulate any immortal essence within the mind; it simply affirms that the mind and the brain are two different entities. Different varieties of this dualism assert that there are different relations between these entities, or none at all, depending on the aspects taken into account by the different currents of study, and depending on whether the variety of dualism refers to substances (the mind is a non-physical entity separated from the body) or to properties (there is no non-physical entity, but the physical world possesses some properties that belong to a distinct class) [2]. For example, *interactionism* considers the mind to be irreducible to the brain while admitting causal links between the two (e.g., [10, 42]), *functionalism* affirms that conscious mental processes are a form of biological adaptation and belong to physical systems of elaboration that are not exclusively human (e.g., [9, 14, 26, 45]), *emergentism* emphasises the creative, systemic aspect of evolutionary changes in the adaptation process (e.g., [34, 52], Sperry,

[60]); *correlationism* presupposes a relationship between mind-brain opposites (e.g., Edelman, [12]; Young, [70]), and so on.

To summarise, we can say that there is an important difference between metaphysical substance dualism and non-metaphysical property dualism: contrary to the former, the latter excludes any immutable, immaterial essence and includes all the distinctions between reductive physicalism and physicalist functionalism, emergentism, interactionism, correlationism, and so on, assuming that all substances are physical but that mental states are related to brain states by many forms of properties. This assumption is based on evidence that shows how mental functions take place within the brain. For example, the clinical neurologist David Weisman ([68], p. 84), in describing the cognitive and personality changes of a patient affected by neurodegenerative pathologies, affirms:

> Yet this folksy, religious, and ancient notion of a soul is mistaken. The evidence that nature provides when natural processes damage the brain does not square with the existence of such a soul. Nature is cruel in countless ways, both human and inhuman. But nature also performs a service if you're willing to take a good, hard look. Nature peels back the skin and shows us exactly what's underneath. What you perceive as your personal unity and control—what you might call your soul—does seem to be true. But the evidence shows that it isn't.

Having eliminated the concept of the metaphysical, essential soul, reductionism seemed to have solved the problem of solipsism, explaining everything as a physical process with various iridescent features. Actually, the problems of intentionality and of empathy—that is, the human ability to understand others' psychological conditions without the use of any neuroimaging technologies (the so-called 'cerebroscope')—remained unresolved until the first decade of the third millennium, when mirror neurons were discovered. This cerebral system provides the physiological mechanism for perception–action coupling, which permits individuals to understand the actions of others and to learn new skills by imitation [49]. By this logic, brains are more or less directly in contact through perceptive activities that support mirroring. Indeed, mirror systems may simulate observed actions and thus internalise inferred mental states. This function seems to particularly assist with language development, as well as the development of emotional competencies and empathy [23]. Furthermore, mirroring might help an individual to understand not only what another person is doing, but also how his/her own self-awareness is developing through constant comparison with others' behaviour and thoughts, as expressed by language (e.g., [65]).

Though mirroring is not possible between human minds and AIS because we do not share any kind of biological mirroring matter, functionalism and emergentism define mental states in terms of functions and occurrences, respectively, meaning that any system that can produce the same pattern of roles will generate the same mental states, including consciousness, regardless of the nature of its physical support. Similarly, the 'machine-state functionalism' of Hilary Putnam [44], which was inspired by Turing machines, affirms that the material constitution of the mind is completely irrelevant with respect to the production of thought and consciousness. This idea is also supported by David J. Chalmers [7], who assumes the 'computational sufficiency' thesis, holding that 'the right kind of computational structure suffices for the possession of a mind, and the thesis of computational explanation, which holds that computation provides a general framework for the explanation of cognitive processes' (p. 323). In this perspective, mental properties are organisationally invariant, whereas the material support may change.

In this knot, one also finds the hidden interest of those who are convinced that it is possible to admit an immortal essence in the dualism of properties. If any mental state is organisationally invariant, then when the brain dies, it would be theoretically conceivable to replace the grey matter with an AIS. This transhumanist perspective is not so fantastical, and it is surely in line with the *Cyborg Manifesto* of materialist posthumanism, which radically confutes any absolute boundary between natural humans and non-natural humans [18]. Cybernetic humanism and posthumanism confute the idea that there must be a necessary link between the specific nature of a material support and its mental/conscious functions. Starting from the concept of 'hybridisation', one can say that humans are no longer natural, since they manipulate their lives with technologies. As the field of augmented reality shows, it is possible to predict that technology will one day enhance people's brain functions, just as mechanical prostheses can replace missing limbs, organs, or perceptive faculties. This idea is clearly featured in the term 'hybrid minds' [8]. This is itself basically in line with the theory of embodied cognition, which considers mind and consciousness to be deeply dependent upon features of the physical body which may be hybridised with technological devices. From this point of view, the standard idea that the mind is identical to, or even realised in, the brain is absolutely insufficient. Largely influenced by the mirror neuron discovery, the embodied cognition theory states that aspects of the agent's body, located largely beyond the brain, play a significant causal or physically constitutive role in all cognitive processing. Mind is not something that happens in or to an individual cranial box, but rather is a social phenomenon constructed by perceptual/imitative action relationships between many bodies [22, 48] and devices, among them AIS. According to this conceptual structure, neuroscience is not the only means to understand the complex phenomenology of social cognition, which recognises the presence of a social consciousness developed through cognitive processes that arise from the interaction between bodies [51], cybernetics is also able to contribute to this understanding through its integration with all branches of the psychology of mind.

4 The Penrose Hypothesis of Consciousness Between Human and Artificial Intelligence

Despite the success of physicalist reductionism, which argues that, somehow, matter causes consciousness (mind, cognition, and so on), the criteria for affirming the identity, similarity, or absolute heterogeneity of two experiences remained to be defined. In fact, it is not clear what 'common matter' between organic and inorganic substrata causes any form of consciousness that can intentionally utilise language to communicate. That is why the FAIR researchers, who are certainly aware of all these issues, killed the two conversing chatbots.

A possible solution appears in Penrose's theory of quantum consciousness. In the 1980s, Penrose suggested that there could be a link between consciousness and quantum mechanics in his book *The Emperor's New Mind*. In Penrose's perspective, regardless of the particular role or mediating effect of consciousness, quantum mechanics is likely to be involved in mind processes because there may be molecular structures in the brain that alter their state in response to a single quantum event. Thanks to Penrose's collaboration with Stuart Hameroff, this idea was further developed into the hypothesis of Orchestrated Objective Reduction (OrchOR), which states that consciousness happens in the brain and originates from a process that takes place inside neurons—or, more particularly, in the microtubules within neurons—rather than in the interaction between neurons (as assumed by conventional neuroscience). In his book *Shadows of the Mind* (1994), Penrose explained that microtubules support quantum superpositions and that the objective collapse of the quantum wavefunction within microtubules is critical for consciousness. However, in his opinion, this quantum collapse is a physical behaviour that transcends the limits of computability, resulting in a non-algorithmic process. As such, the fundamental difference between human consciousness and a putative AIS consciousness is that the first one is not computational. This implies that human consciousness is not as mouldable as any conventional Turing machine and that the human mind has abilities that no AIS could possess because of the non-computable physics of the OrchOR mechanism (see also [17, 38]).

Despite other projects that examine and confirm the correlation (if not the causality) between quantum processes and consciousness (e.g., [28, 40]), this important insight is qualified by two further problems. The first derives from the fact that human consciousness/mind and brain functioning may also be explained in a modular way (see [15]). The second pertains to the fact that human-like brain dynamics can emerge in AIS circuits. Indeed, according to the perspective of AI's third wave, Turing machines can assume the quantum computing model—that is, the use of quantum–mechanical phenomena such as superposition and entanglement to perform computation [31]. This results in a possible quantum Turing machine. Penrose's limiting judgment regarding Turing machines (that their computational cognition differs from human cognition) applies only to AI's first (handcrafted knowledge to describe) and second (statistical learning to categorise) waves. Indeed, the third wave intends to develop the contextual adaptations necessary to explain facts and creatively solve derived problems. For example, in the DARPA perspective of John Launchbury [25], the next steps for AI include the ability to abstract, to generate new meaning, and to reason, plan, and decide, which neural networks are already partially able to do (learning/teaching machines and artificial super intelligence).

If consciousness (either artificial or human) is a function of and not identical to the brain, then quantum mechanics could be the common feature of organic and inorganic matter that causes mind and consciousness. If so, we cannot say that human consciousness is only human because of quantum mechanics, as consciousness may appear in both organic and inorganic matter precisely because of quantum mechanics. Quantum mechanics may serve as a medium between the Cartesian *res extensa* (organic and inorganic material substrate) and *res cogitans* (mind, consciousness, feeling, experience). Indeed, Penrose's microtubules are similar to Descartes' pineal gland. Quantum mechanics could also explain what Descartes was unable to, providing a definitive material foundation for the dualistic concepts of function, emergence, parallelism, and so on, as related to the concepts of consciousness and mind. However, where the Cartesian solution guaranteed the existence of an essential soul, quantum processes seem to be unable to do the same. Let us say that it is possible, even for a moment, to transfer a quantum consciousness function from a natural material substrate to an artificial one. Regardless, Penrose argues that everything is bound to end because of the cyclicity of the universe, which implies the phasic dismantling of everything—the so-called 'Conformal Cyclic Cosmology' (CCC) [19, 37, 39].

All this means that, even if it is theoretically possible to solve the problem of matter's relationship with consciousness, this kind of consciousness is inescapably mortal from a reductionist point of view. This means that nothing survives annihilation. With respect to the death issue, the contrast between metaphysical (Platonic/Cartesian) and reductionist substance dualism consists in the representation of death as a passage (as conceived by the former) or as total annihilation (as conceived by the latter). All scientific knowledge states, a priori, that total annihilation is realistic. No matter and no consciousness or soul survives this absolute annihilation, whether the inner universe or the life of every single inhabitant of Earth. Every single being—humans, animals, AIS—is destined for annihilation, just as the universe is. In this conviction lies the most important epistemological problem: scientists assume a priori that it is true that things 'become'—that they emerge from and return to nothing. As Thomas Nagel ([34], p. 7) stated: 'It is true that both the time before a man's birth and the time after his death are times when he does not exist'. This is the problem that Emanuele Severino solves by proving that annihilation is impossible. Nothing comes from or returns to 'nihil', neither quanta nor the most complex forms of life and galaxies.

5   Severino's Ontological Solution

Because the concept of 'reality'—glossed as physicalist reductionism's alternative solution to hyperbolic doubt, in place of the transcendent, Cartesian God—can be interpreted in various ways, particularly in terms of its relation to thought and the thinking subject, Emanuele Severino instead relies on the concept of 'truth', which provides the only secure basis for intersubjectivity. Truth is, by necessity, an incontrovertible discourse. In Severino's opinion, philosophy has robbed myth of credibility, starting with the pre-Socratics, and Parmenides, in particular. The Eleatics claimed that truth is the opposite of myth, and that the latter does not guarantee incontrovertible knowledge. Furthermore, truth is the necessary discourse inherent to 'being' (as 'einai' - εἶναι) versus 'non-being/nothing' (as 'me ón' - μὴ ὀν). In the first iteration of this ontology, truth indicates the knowledge 'that it is and that it is not possible for it not to be [...] that it is not and that it must not be' (*On Nature*, Frag. 2); in other words: 'It is necessary to say and to think that what is is; for it is to be,/ but nothing it is not' (*On Nature*, Frag. 6). Such a 'being' necessarily is, and 'being' cannot be 'nihil'. Absolute Being does not change, and the discourse of truth asserts that reality is, and must necessarily be, unity, and thus totally identical to itself. In the 'well-rounded reality', any change is impossible: there is no before or after; 'becoming' (transformation, change, alteration) is impossible and illusory. Persuasive truth consists in any statement affirming the necessity of Being and confuting the 'opinions of mortals' (myths), whose arguments are not truly warranted. This is the beginning of ontology and its relationship with truth, from which the concept of reality arises as the science of being, of existence, and of all that there is [61].

As Severino discusses throughout his oeuvre, the concept of truth sprang from Parmenides' ontology and founded all of Western thought (e.g., [55–57]) on the basis of three logical axes: (1) the *Principle of Identity*: $A \equiv A$ (every being is identical with itself), or '$(\forall x) (x = x)$' (in which $\forall$ means 'for all'), or, simply, 'x is x'; (2) the *Principle of Non-Contradiction*: for all propositions p, it is impossible for both p and not p to be true; and (3) the *Principle of Excluded Middle/Third*: there is no third or middle 'true' proposition between them. According to Severino, this fundamental tripartite system was originally developed by metaphysics. However, Severino shows that, in Parmenides' formulation, any transformation is illusive and opposite to the 'path of truth'. The conviction that Being constitutionally changes is the mythical 'way of mortals', which is characterised by faith in the phenomenon of becoming and by the contradictory opinions deriving from experience. In his analysis of the history of Western thought, Severino [58] affirms that the first attempt to remedy the negative relationship between phenomena and worldly appearances was realised by Plato, who, in the *Sophist*, introduced the notion of 'relative non-being/nothing' and defined the 'multiplicity of beings' as the 'énantíon' (εναντίον), which is the oscillation of everything between Being (τὸ ὄν) and non-being/nothing (μὴ ὄν). Change consists of this oscillation of all concrete (material) things between being and nothing. In order to retain the logic of Parmenides' original thought, Aristotle systematised the difference between metaphysical, or absolute and immutable, Being (God) as the first and final cause, on the one hand, and physical beings—contingent entities oscillating between being and nothing and subjected to the laws of time and space—on the other. Platonic and Aristotelian metaphysics tried to provide the most secure definition of truth: a relationship between absolute Being (God) and contingent beings wherein the former determines/causes the destiny of the latter and grants them eternity after death. Eternal and immutable essences shape matter, transforming it into substances, and as such it is possible to recognise the form of something by looking at it because the soul accesses the knowledge of these universal essences.

In Severino's thought, the fourth and most important principle that founds true knowledge was partially formulated by Aristotle: the *elenchos* (ἔλεγχος). In the dialogue of the *pólemos* (πόλεμος) between the discourse of truth, which respects the tripartite logical system, and its denier, the *elenchic* argumentation of truth demonstrates the auto-contradiction of the denier's discourse, which implicitly assumes truth as the basis of its negation. Those who deny truth effectively deny any basis for their own confutation. Thus, the *elenchos* is the fourth fundamental basis of logical argumentation that metaphysics enshrined as the 'first philosophy' [55]. This metaphysical logic structure was assumed by the monotheistic Abrahamic religions, which attempted to rationally demonstrate the necessity of God (first and last cause of all material things, which oscillate between being and nothing) and of the human afterlife in a true and non-mythological way.

In Severino's view, this solution was not only the greatest attempt to solve the existential question, but also, simultaneously, the most important failure to unequivocally clarify the relationship between what appears and what truly is (reality). In his opinion, although metaphysics pointed out the four principles for the first time, sought true knowledge to defy the rhetoric of death as total annihilation and thus mitigated people's fear of death by grounding the idea of transcendence in logic, in practice it betrayed this goal. Ultimately, metaphysics was unable to achieve unquestionable, or true, knowledge.

Traditional Western thought sought to differentiate between indubitable and mythological remedies, considering the former as true/logical discourse and the latter as a matter of opinions and illusions. Nevertheless, Severino argues (1982/2015) that positivism, materialism, and reductionism irreversibly confuted metaphysics because they understood that, if a little thing can arise from nothing and return to nothing, then everything—including Being and the universe itself—can do the same. As such, the notion of God is inadequate—and ultimately unnecessary—to explain physical laws relating to becoming (Occam's razor). If we admit the phenomenon of becoming in a partial way (as in the oscillation between being and nothing), we can also admit it in a total way, wherein everything arises from and returns to nothing, as the Big Bang and Big Crunch theories describe. All this means that neither soul nor quantum consciousness can retain any subjective identity following annihilation. Every being springs from and returns to non-being/nothing. But this is contemporary thought; if, on the one hand, it correctly and irrevocably refutes metaphysics, on the other hand it shares with metaphysics the same fundamental error, but expresses it in an even more radical way. This ontological representation of becoming may be taken as 'real', but not as 'true', as it directly contradicts the concept of truth by assuming, a priori, that 'being is nothing'. As Severino explicitly explains, because being cannot be nothing, asserting that what 'is not yet' and what 'is no longer' is nothing is tantamount to saying that what is nothing. In Severino's perspective, this is the fundamental contradiction of 'nihilism': affirming that there is a time in which being is no longer or not yet (past and future) is the same as stating that 'being is nothing'. And because this does not respect any logical rule, it is an absolute and radical contradiction. This is the basic error of both science and metaphysics.

A similar critique was intuited by some immanentist philosophers, among them Giordano Bruno and Spinoza, but they did not criticise metaphysics in a systematic way. Some eternalist philosophers also tended to argue in this direction, taking positions very close to Einstein in an attempt to eliminate the time dimension. Indeed, recent epistemologies, which assume the contraposition between different conceptualisations of time [61], are based on Parmenides' reflections on being, as re-actualised by the philosopher John McTaggart [29]. McTaggart affirmed the unreality of time in the same period as Einstein's general theory of relativity and Gödel's incompleteness theorems, thus resurrecting discussions about eternity. According to [29], there are two distinct modes in which all events can be ordered in time: the A-series and the B-series. The logical and linguistic expression of the two series, from which specific theories of time derive, are radically different because the former is tensed and the latter is tenseless. Whereas the A-theory of time asserts that the series of temporal positions are in continual transformation, the B-theory, through the concept of 'endurance', updates Parmenides' perspective, arguing that the flow of time is an illusion because past, present, and future are equally real. There are two principal varieties of the A-theory: presentism and the growing block universe. The first holds that only present objects exist, or, more precisely, that it is always true that only present objects exist [5], whereas the second contends that both present and past objects exist, but not future ones [65]. Instead, the B-theory belongs to eternalism, which utilises the B-series, according to which temporal collocation is a quality of everything, and time is akin to the dimensions of space. Indeed, it affirms that persistence through time is similar to extension through space; thus, any being that exists in time has temporal 'extension' in the various sub-regions of the total area of time [1, 47]. The argument between A-theory and B-theory has been developed in the field of analytic philosophy to inform the more significant new epistemologies (e.g., [27, 43]), and it entails the argument of 4-dimensionalism versus 3-dimensionalism (3D) that emerged via the theory of relativity [20].

The logical languages assumed by the A-theory and the B-theory are necessarily different because, in the tensed view of semantics, propositions have truth values 'at a given time' rather than having truth values simpliciter. In contrast, in the tenseless view, propositions have truth values simpliciter rather than having truth values at a given time. However, it appears that this argument is endless, and it will be, for some time, unsolvable. As such, any aprioristic perspective assumed ideologically by scientists is justified. Nonetheless, Emanuele Severino [55] solved the contraposition between eternalism and temporalism. Although the current dispute between eternalists and presentists may create the impression that the problem of eternity and the ensuing questions concerning the afterlife are pointless, Severino shows how the truth discourse on so-called "reality" is the only definition of reality that maintains an immediate correspondence, or identity, between what appears phenomenologically and the logical structure of what is. In other words, a discourse is only 'true' if it shows, in an incontrovertible way (*elenchos*), that what appears is identical to and recognisable as itself (identity principle), that it cannot be another thing or become nothing (non-contradiction principle), and that any intermediate position is impossible (excluded middle). Anything that appears is necessarily eternal (exists ontologically), even when it does not yet appear and no longer appears. The becoming of the spectacle of matter is a phenomenological fact, and this phenomenon is something as well. It is not nothing. Actually, consciousness consists in the appearing of the phenomena itself, which is the spectacle of being, that eternally appears in the infinite being, not limited by nothingness.

Severino presents an incontrovertible, unitary, and structurally coherent system of thought and a vigorous critique of the nihilism of Western thought, both traditional (metaphysics) and contemporary (neo-positivism, materialism, and so on), for its inability to maintain the fundamental ontological and logical structure of truth. From this perspective, Parmenides was the beginning of the 'path of night' (error, nihilism), wherein Western philosophy achieved its full coherence by denying truth, as contemporary thought does with scientific epistemology, but in a way that evokes the most profound alienation. Nihilism—that is, the path of error, to which both metaphysics and contemporary thought belong—is the alienation of authentic truth. Truth is the knowledge that reveals the self-contradiction and self-negation of this alienating error, and thus confirms the necessary self-negation of metaphysics and contemporary thought. Severino restores the authentic, incontrovertible discourse of truth, which is crucially different from the 'reality' proposed by traditional thought or any contemporary or scientific perspectives. In essence, he asserts that every being is eternal, meaning that (1) everything that is not nothing is a being, (2) it is necessary for each being to be, and, more specifically, to be as it is, and (3) it is impossible for any being to not be at any time.

Any type of being appearing in the present is both real and eternal; moreover, time is the phenomenal appearing of eternal beings that go into and come out of the 'circle of appearing'. Because any annihilation is impossible, this 'transformation' is simply the appearance of subsequence—of being's entrance into and departure from so-called 'reality', which may be defined as the 'limited circle of appearing'. This point is crucial: there is no contradiction between the appearance of these sequences, or of qualities in the flow of time, and the eternity of what appears. The nihilist error lies in equating this process of continual (re)appearing (restated as 'becoming') in the world with a process of creation–annihilation, wherein death amounts to annihilation. Severino's thought reveals the nihilism of traditional and contemporary thought and solves their fundamental and essential contradiction. Contrary to what Western thought aprioristically assumes, there is no becoming, that is, it is impossible to come *ex nihilo* or depart *ad nihilum*. Indeed, the content that actually appears in any consciousness does not attest in any way to the creation or annihilation of beings.

6   Conclusions

> An old king is dying. A sword has been driven deep inside his breast. All around the king, friends, foes, courtiers, jesters dance: each believes himself to be the one who drove the sword into the king's breast. And yet there the sword stands, plunged in that breast, regardless of the will of all those reckless dancers. The dying king is philosophy (in its strong, Greek meaning, namely metaphysical *epistéme*). The reckless dancers are all the criticism that has been addressed to metaphysics throughout the history of philosophy. No part of this criticism can really kill philosophy. Yet, philosophy really is dying. Of an illness that our culture still cannot identify. It is dying under the gaze of destiny, of which philosophy as *epistéme* is the deformed image.

This excerpt, taken from Severino's *Studi di filosofia della prassi* (p. 396–7), finally reintroduces us to the topic that is most difficult to address in contemporary Western culture due to the angst deriving from technoscience's redefinition of the boundaries between life and death. The matter is relevant as it concerns our representations of death, the terror generated by representations of death as absolute annihilation, and the need to soothe the anticipatory angst of consciousness's obliteration.

Contemporary secularism assumes that the end of transcendence is the beginning of truth. However, we live in a culture where religions continue to promote the truthful metaphysical idea of existence after death. Thanks to its consolatory power, this idea has broad influence over common sense, even while coexisting with secularism, which declares any negation of absolute annihilation illusory. The present historical period is very emotionally fraught because we know that the greatest attempt to save humans from absolute annihilation—that is, the metaphysical argument—has failed. Western people are attracted by messages promising salvation from death, which reinforce the primeval hope that religious thought can be essentially reliable. However, these same people are exposed to the dramatic refutation of these promises by scientific and academic knowledge, which are based on reductionist epistemologies.

Our discussion has evidenced how the complexity of the epistemological argument inherent to the meaning of soul/mind/consciousness and its death does not allow for any simplification. It may be that the mind–brain problem is one example of a biased and growing tendency to generalise specific models deriving from sectorial and empirical lines of reasoning, which should remain inscribed in limited areas, especially with respect to questions of death and dying. Nowadays, the fear of death, regarded as absolute annihilation, is a real source of profound anxiety. Despite the fact that contemporary people are more or less aware of their 'wishful thinking', convinced that science is a more credible source of knowledge than religion, and thus that everything comes from nothing (Big Bang) and returns to nothing (Big Crunch), they nonetheless perceive themselves as eternal. Because of this, they are constantly looking for some reasonably possible, coherent framework to consolidate their feelings into something that language can represent. Indeed, religious mythologies maintain the scope for possibility, confirming that, whatever they may think they know, what is really knowable remains a mystery. And mysteries, as Wittgenstein himself said, cannot be spoken of. Moreover, people's faith in this mystery is reinforced by the fact that science itself declares that it does not know everything that there is to know, and even that it is fallible.

However, Severino solves the mystery beyond the point of faith: we correctly perceive eternity despite the nihilistic language of our daily lives. Severino's transcendental argument may help to renew the Platonic idea that knowledge is memory (Meno, 79e–82b), as, in the eternalist perspective, this idea could explain why we believe in an afterlife. If our consciousness already lives/lived in another dimension in the eternal and infinite universe of Being, then knowledge may surface in our unconscious as memories.

The epistemological discussion could proceed, but we must conclude here and return to the two chatbots. In Severino's incontrovertible indication of truth, consciousness is the phenomenological appearing of everything, which is as eternal as any matter and is identical to itself, so it cannot be reduced to matter. The relationships between consciousness (soul, spirit, mind, and so on) and matter (grey or otherwise) cannot be reduced to their reciprocal identity.

Quantum mechanics may cause consciousness to emerge in AIS, but it may not wait for humans to produce it in a laboratory by constructing a quantum Turing computer. If consciousness (the appearing) is not restricted to a brain cavity, then we must recognise that it transcends the individual and human dimension, and thus humans' ability to recognise its presence. This difficulty is necessarily linked to the nihilistic language that characterises all the studies on this problem. Severino's argument can help us recognise that, whenever something appears, a specific dimension of consciousness is involved. This means that the quantum computer is probably already autogenerating from the most evolved computational AIS, because a specific form of consciousness is even now appearing in this mechanical being, hidden from human awareness. It is really 'wishful thinking' to believe that reality is limited to what humans understand. Though we are not yet able to consciously construct a quantum Turing machine, it is not so improbable that the two killed chatbots were already more or less intentionally improving and developing their own quantum consciousness.

References

1. Baldwin, T. (1999). Back to the present. *Philosophy, 74*(2), 177–197. https://doi.org/10.1017/s003181919900025x
2. Bechtel, W. (1988). *Philosophy of mind*. Lawrence Erlbaum.
3. Berchman, R., & Finamore, J. (2005). *Studies in platonism, neoplatonism, and the platonic tradition*. Brill. https://brill.com/view/serial/SPNP
4. Bianco, S., Testoni, I., Palmieri, A., Solomon, S., & Hart, J. (2019). The psychological correlates of decreased death anxiety after a near-death experience: The role of self-esteem, mindfulness, and death representations. *Journal of Humanistic Psychology*. https://doi.org/10.1177/0022167819892107
5. Bourne, C. (2006). *A future for presentism*. Oxford University Press.
6. Carnap, R. (1967). *The logical structure of the world and pseudoproblems in philosophy*. University of California Press.
7. Chalmers, D. J. (2011). A computational foundation for the study of cognition. *Journal of Cognitive Science, 12*(4), 325–359. https://doi.org/10.17791/jcs.2011.12.4.325
8. Clark, A. (2003). *Natural-Born cyborgs: Minds, technologies, and the future of human intelligence*. Oxford University Press.
9. Dennett, D. (1991). *Consciousness explained*. Brown.

10. Eccles, J. (1973). *The understanding of the brain*. McGraw-Hill.
11. Elkaisy-Friemuth, M., & Dillon, J. M. (2009). *The afterlife of the platonic soul. Reflections of platonic psychology in the monotheistic religions*. Brill.
12. Edelman, G. M. (1992). *Bright air, brilliant fire: On the matter of the mind*. New York: Basic Books.
13. Field, M. (2017). Facebook shuts down robots after they invent their own language. *The Telegraph*. https://www.telegraph.co.uk/technology/2017/08/01/facebook-shuts-robots-invent-language/
14. Fodor, J. A. (1980). Methodological solipsism considered as a research strategy in cognitive psychology. *Behavioral and Brain Sciences, 3*(1), 63–73. https://doi.org/10.1017/s0140525x00001771
15. Gazzaniga, M. S. (1985). *Social brain*. Basic Books.
16. Goldenberg, J. L., & Arndt, J. (2008). The implications of death for health: A terror management health model for behavioral health promotion. *Psychological Review, 115*(4), 1032–1053. https://doi.org/10.1037/a0013326
17. Hameroff, S., & Penrose, R. (2014). Consciousness in the universe: A review of the 'Orch OR' theory. *Physics of Life Reviews, 11*(1), 1–40. https://doi.org/10.1016/j.plrev.2013.08.002
18. Haraway, D. (1990). *Simians, cyborgs, and women: The reinvention of nature* (1st edn). Routledge.
19. Hawking, S., & Penrose, R. (1996). *The nature of space and time*. Princeton University Press.
20. Hawley, K. (2001). *How things persist*. Oxford University Press.
21. Hines, T. (2015). Brain, language, and survival after death. In: M. Martin, & K. Augustine (eds.), *The myth of an afterlife: The case against life after death* (pp. 183–193). Rowman & Littlefield Publishers.
22. Iacoboni, M., & Dapretto, M. (2006). The mirror neuron system and the consequences of its dysfunction. *Nature Reviews Neuroscience, 7*(12), 942–951. https://doi.org/10.1038/nrn2024

23. Iacoboni, M., Molnar-Szakacs, I., Gallese, V., Buccino, G., Mazziotta, J. C., & Rizzolatti, G. (2005). Grasping the intentions of others with one's own mirror neuron system. *PLoS Biology, 3*(3), e79. https://doi.org/10.1371/journal.pbio.0030079
24. Kounios, J., & Beeman, M. (2009). The Aha! moment. *Current Directions in Psychological Science, 18*(4), 210–216. https://doi.org/10.1111/j.1467-8721.2009.01638.x
25. Launchbury, J. (2018) [Slides from the PowerPoint talk "A DARPA perspective on artificial intelligence"]. https://www.darpa.mil/attachments/AIFull.pdf
26. Levin, J. (2008). Taking type-B materialism seriously. *Mind & Language, 23*(4), 402–425. https://doi.org/10.1111/j.1468-0017.2008.00349.x
27. Maudlin, T., & Oxford University Press. (2007). *The metaphysics within physics*. Amsterdam University Press.
28. McFadden, J. (2007). Conscious electromagnetic (CEMI) field theory. *NeuroQuantology, 5*(3), 262–270. https://doi.org/10.14704/nq.2007.5.3.135
29. McTaggart, J. E. (1908). The unreality of time. *Mind, XVI, I*(4), 457–474. https://doi.org/10.1093/mind/xvii.4.457
30. Miller, A., Feng, W., Fisch, A., Lu, J., Batra, D., Bordes., Parikh, D., & Weston, J. (2017). ParlAI: A dialog research software platform. In: *Proceedings of the 2017 conference on empirical methods in natural language processing: System demonstrations* (pp. 79–84). Denmark Association for Computational Linguistics.
31. Molina, A., & Watrous, J. (2019). Revisiting the simulation of quantum Turing machines by quantum circuits. *Proceedings of the Royal Society A: Mathematical, Physical and Engineering Sciences, 475*(2226), 20180767. https://doi.org/10.1098/rspa.2018.0767
32. Moreman, C. M. (2008). *Beyond the threshold: Afterlife beliefs and experiences in world religions*. Rowman & Littlefield Publishers.

33. Nagasawa, Y., & Matheson, B. (2017). *The Palgrave handbook of the afterlife (Palgrave Frontiers in Philosophy of Religion)*. Palgrave Macmillan.
34. Nagel, T. (1979). *Mortal questions*. Cambridge University Press.
35. Neisser, U. (1976). *Cognition and reality: Principles and implications of cognitive psychology*. W. H. Freeman and Company.
36. Penrose, R. (1989). *The emperor's new mind—concerning computers, minds, and the laws of physics*. Oxford University Press.
37. Penrose, R. (2014). On the gravitization of quantum mechanics 2: Conformal cyclic cosmology. *Foundations of Physics, 44*, 873–890. https://doi.org/10.1007/s10701-013-9763-z
38. Penrose, R. (1994). *Shadows of the mind: A search for the missing science of consciousness*. Oxford University Press.
39. Penrose, R. (2010). *Cycles of time: An extraordinary new view of the universe*. Bodley Head.
40. Pereira, A., Jr. (2007). Astrocyte-trapped calcium ions: The hypothesis of a quantum-like conscious protectorate. *Quantum Biosystems, 2*, 80–92.
41. Piccinini, G., & Bahar, S. (2015). No mental life after brain death: The argument from the neural localization of mental functions. In: M. Martin, & K. Augustine (eds.), *The myth of an afterlife: The case against life after death* (pp. 135–170). Rowman & Littlefield Publishers.
42. Popper, K., & Eccles, J. C. (1977). *The self and its brain*. Routledge.
43. Prior, A. N. (1967). *Past*. Oxford University Press.
44. Putnam, H. (1975). *Philosophical papers volume 2: Mind, language and reality*. Cambridge University Press.
45. Putnam, H. (1988). *Representation and reality*. Mit Pr.
46. Pyszczynski, T., Solomon, S., & Greenberg, J. (2015). Thirty years of terror management theory. *Advances in Experimental Social Psychology*, 1–70. https://doi.org/10.1016/bs.aesp.2015.03.001

47. Richard, M. (1982). Temporalism and eternalism. *Philosophical Studies, 39*(1), 1–13. https://doi.org/10.1007/bf00354808
48. Rizzolatti, G., & Craighero, L. (2004). The mirror-neuron system. *Annual Review of Neuroscience, 27*(1), 169–192. https://doi.org/10.1146/annurev.neuro.27.070203.144230
49. Rizzolatti, G., & Fogassi, L. (2014). The mirror mechanism: Recent findings and perspectives. *Philosophical Transactions of the Royal Society B: Biological Sciences, 369*(1644), 20130420. https://doi.org/10.1098/rstb.2013.0420
50. Schank, R. C. (1972). Conceptual dependency: A theory of natural language understanding. *Cognitive Psychology, 3*(4), 552–631. https://doi.org/10.1016/0010-0285(72)90022-9
51. Schütz-Bosbach, S., & Prinz, W. (2007). Perceptual resonance: Action-induced modulation of perception. *Trends in Cognitive Sciences, 11*(8), 349–355. https://doi.org/10.1016/j.tics.2007.06.005
52. Searle, J. R. (1984). *Minds, brains and science (1984 Reith Lectures)*. Harvard University Press.
53. Segal, A. (2004). *Life after death: A history of the afterlife in western religion*. Image.
54. Severino, E. (1979). *La struttura originaria*. Adelphi.
55. Severino, E. (1982). *Essenza del Nichilismo*. Adelphi.
56. Severino, E. (1984). *Studi di filosofia della prassi*. Adelphi.
57. Severino, E. (1985). *Il parricidio mancato*. Adelphi.
58. Solomon, S., Testoni, I., & Bianco, S. (2017). Clash of civilizations? Terror management theory and the role of the ontological representations of death in contemporary global crisis. *TPM: Testing, Psychometrics, Methodology in Applied Psychology, 24*, 379–398. https://doi.org/10.4473/TPM24.3.5
59. Solomon, S., Greenberg, J., & Pyszczynski, T. (2012). *The worm at the core: On the role of death in life*. Random House.
60. Sperry, R. W. (1986). Consciousness, personal identity and the divided brain. In: F. Lepore, M. Ptito, & H. Jasper

(eds.), *Two Hemispheres-One Brain: Functions of the Corpus Callosum* (pp. 3-20). NewYork: Alan R. Liss, Inc.
61. Stoneham, T. (2009). Time and truth: The presentism-eternalism debate. *Philosophy, 84*(2), 201–218. https://doi.org/10.1017/s0031819109000187
62. Testoni, I., Ancona, D., & Ronconi, L. (2015). The ontological representation of death. *OMEGA—Journal of Death and Dying, 71*(1), 60–81. https://doi.org/10.1177/0030222814568289
63. Testoni, I., Facco, E., & Perelda, F. (2017). Toward a new eternalist paradigm for afterlife studies: The case of the near-death experiences argument. *World Futures, 73*(7), 442–456. https://doi.org/10.1080/02604027.2017.1357935
64. Thompson, T. (2008). Self-awareness: Behavior analysis and neuroscience. *The Behavior Analyst, 31*(2), 137–144. https://doi.org/10.1007/bf03392167
65. Tooley, M. (1997). *Time, tense, and causation* (1st edn). Oxford University Press.
66. Turing, A. M. (1950). Computing machinery and intelligence. *Mind, LIX*(236), 433–460. https://doi.org/10.1093/mind/lix.236.433
67. Walter, T. (1996). *The eclipse of eternity*. Palgrave Macmillan.
68. Weisman, D. (2015). Dissolution into death: The mind's last symptoms indicate annihilation. In: M. Martin, & K. Augustine (eds.), *The myth of an afterlife: The case against life after death* (pp. 83–104). Rowman & Littlefield Publishers.
69. Wittgenstein, L. (1922). *Tractatus logico-philosophicus*. Routledge & Kegan.
70. Young, J. Z. (1987). *Philosophy and the brain*. Oxford University Press.

The Brain Is not a Stupid Star

Giuseppe Vitiello

Abstract The activity of the neocortex presents the formation of extended configurations of oscillatory motions modulated in amplitude and phase and involving myriads of neurons. As observed by Lashley, nerve impulses are transmitted from cell to cell through defined cell connections. However, all behavior seems to be determined by masses of excitations, within general fields of activity, without reference to particular nerve cells. Freeman has stressed the role played by chaos underlying the ability of the brain to respond flexibly to the outside world. These observations support the remark, attributed to Aristotle, that the brain is not a stupid star, which in its perennial

---

G. Vitiello (✉)
Dipartimento Di Fisica "E.R. Caianiello", Università Di Salerno, Via Giovanni Paolo II, 132, 80084 Fisciano (Salerno), Italia
e-mail: vitiello@sa.infn.it

trajectory passes always through the same point in a fully predictable way. On the contrary, brains appear to proceed by steps that do not necessarily belong to a strictly predictive chain of steps, but behave like a 'machine making mistakes', intrinsic erratic devices. These features of neural dynamics are discussed within the framework of the dissipative quantum model of the brain and with reference to AI systems and research programs. If it is ever possible to build a device endowed with consciousness, it must possess unpredictability of behavior, infidelity, and inalienable freedom; and must be called *Spartacus*.

1 Introduction

I have already used the title of this paper "The brain is not a stupid star" as the title of a section of my contribution to the book by Robert Kozma and Walter Freeman on cognitive phase transitions in the cerebral cortex [1]. I mentioned there that, according to Aristotle, stars have a stupid behavior since in their perennial trajectories they always pass through the same point in a fully predictable way. I do not know if such an observation was really made by Aristotle. However, the suggested idea is that the brain is not a stupid star since its behavior is not fully predictable. The point is that stupidity (non-intelligence?) is associated with fully predictable behavior within broadly unchanged boundary conditions. The brain, on the contrary, is motivated in its behavior by intentional tasks, which although partially conditioned by the environment in which it is embedded, are definitely formulated and pursued by the brain in its action–perception cycle. As stressed by Pribram [2–4], in brain activity there is always a content of 'attention' in perception and of 'intention' in action. Neuronal activity, according to Freeman, acts "as a unified whole in shaping each intentional action at each moment" [5]. In pursuing our best to-be-in-the-world, we are indeed guided by our changeable volition and intention [6, 7].

These observations might lead straight to the theme of artificial intelligence (AI), with reference to one of the possible meanings of "intelligence" and the predictability of the functioning of a device. However, let me discuss first some of the features of the brain's functional activity as described within the dissipative quantum model of brain [8, 9]. In Sect. 2 some general features in the brain studies are briefly presented. Section 3 is devoted to some notions of quantum field theory (QFT), in particular to coherence, a basic ingredient in the many-body model of the brain and the dissipative quantum model of the brain, which will be introduced in Sects. 4 and 5, respectively. In Sect. 7, I discuss some features of AI in connection with the chaotic classical trajectories in the space of memory described by the dissipative model and introduced in Sect. 6. Section 8 is devoted to concluding remarks.

2 Lashley's Dilemma

Let me start by mentioning that Karl Lashley, in commenting the experiments he conducted in the 1940s, proposed the following dilemma to neuroscience scholars [10, pp. 302–306]:

> Here is the dilemma. Nerve impulses are transmitted [...] from cell to cell through defined cell connections. Yet all behavior seems to be determined by masses of excitations [...] within general fields of activity, without reference to particular nerve cells [...]. What kind of nervous organization can ever account for patterns of excitations without

well-defined and specialized channels of cellular communication? The problem is almost universal in the activity of the nervous system.

Lashley thus arrived at the formulation of the hypothesis of mass action in brain activity. His experimental observations, fully confirmed by subsequent studies (see, for example, [11]), led Karl Pribram in the 1960s to propose the hypothesis that for the brain one could speak of coherence, a central notion in quantum optics, and use the metaphor of the hologram [2, 3]. One of the characteristic features of a hologram is that knowledge of a detail in any point of the image allows the reconstruction of the whole image. Such a possibility comes from the fact that the photons, the quanta of the electromagnetic field, which constitute the laser used to produce and read the hologram, oscillate in phase. The laser is made by *coherently* oscillating photons. The laser is what results from the harmony, if I am allowed to use this term, of the *coherence* of photons. Natural light ('non-laser' light) is instead made of photons that are not in phase, i.e., they are not coherent. Beyond the specificity of the model proposed by Pribram, the value of his intuition consists in the fact that for brain activity we can speak of coherence.

The coherence hypothesis for the brain's functional activity is actually confirmed by the observation of widespread cooperation between a huge number of neurons over vast brain areas. Analysis of the potentials measured with the electroencephalogram (EEG) and with the magnetoencephalogram (MEG) shows that the neural activity of the neocortex presents the formation of extended configurations of oscillatory motions modulated in amplitude (AM) and phase (PM) [12, 13]. These configurations extend over almost the entire cerebral hemisphere for rabbits and cats, on domains of linear dimensions up to twenty centimeters in the human brain, and have almost zero phase dispersion [14, 15]. The associated oscillation frequencies are in the 12–80 Hz range (the so-called beta and gamma waves). The patterns of neuronal oscillations dissolve in a few tens of milliseconds and others appear in different configurations, with frequencies in the 3–12 Hz range (theta and alpha waves) [12, 16–21].

A huge number of cells and other biologically active units enter into brain activity. For example, a weak olfactory stimulus activates about a thousand neurons in the olfactory bulb that produce the excitation of a million neurons and the inhibitory activity of 10 million neurons that propagates in 5–10 ms over a distance of about 10 mm, although the average axon lengths are about 1 mm and the synaptic propagation times are about 10 times longer [22]. It is evident that in the presence of such huge numbers and such complexity, the study of brain functions cannot be limited to the knowledge of the properties of the individual elementary components. This is certainly necessary, but it is by no means sufficient. Ricciardi and Umezawa write [23]:

> [...] in the case of natural brains, it might be pure optimism to hope to determine the numerical values for the coupling coefficients and the thresholds of all neurons by means of anatomical or physiological methods. [...] First of all, at which level should the brain be studied and described? In other words, is it essential to know the behavior in time of any single neuron in order to understand the behavior of natural brains? Probably the answer is negative. The behavior of any single neuron should not be significant for the functioning of the whole brain, otherwise a higher and higher degree of malfunctioning should be observed, unless to assume the existence of "special" neurons, characterized by an exceptionally long half life: or to postulate

a huge redundancy in the circuitry of the brain. However, to our knowledge, there has been no evidence which shows the existence of such "special" neurons, and to invoke the redundancy is not the best way to answer the question.

Referring to biological systems in general, but his words might be applied equally well to the brain, Schrödinger observes that [24, p. 79],

> [...] it needs no poetical imagination but only clear and sober scientific reflection to recognize that we are here obviously faced with events whose regular and lawful unfolding is guided by a "mechanism" entirely different from the "probability mechanism" of physics.

The discovery of the constituents of biological systems and the knowledge of their specific properties certainly constitutes a great success of molecular biology. The problem now is one of understanding how to put these elementary constituents together in such a way as to generate the complex macroscopic behavior of the system [25–30]. As observed elsewhere [31],

> In very general terms, the problem is the one of the transition from naturalism, that is, from the knowledge of the catalogs of elementary components, to understanding the dynamics that accounts for the relationships that bind these components and describes the behavior of the system as a whole. The phase of naturalism is obviously essential and requires an enormous effort of careful and patient investigation. Although it is necessary, it is not sufficient for the purposes of a full understanding of the phenomena that are the object of our study. Knowing is not yet understanding.

According to Schrödinger, in the study of living matter, the distinction has to be made between the *two ways of producing orderliness*: ordering generated by the "statistical mechanisms" and ordering generated by "dynamical" interactions [24, p. 80].

We might conclude that Herbert Fröhlich [25, 26], Umezawa and Ricciardi [23, 32, 33], Karl Pribram [2–4], and Walter Freeman [22, 34], have each in their own way shown how, by focusing on "masses of excitation" and "fields of activity", in Lashley's words, naturalism may become Galilean science (see [34, 35]).

The notion of coherence and the associated mathematical formalism provided by QFT have proved to be formidable tools in the study of biological systems in general [27–30] and of the brain in particular [8, 9, 18, 36]. Before beginning the discussion of brain modeling, I will therefore introduce in the next section a few general notions of QFT.

3 Coherence: From the Microscopic to the Macroscopic

The concept of coherence is central to quantum physics, where it allows us to explain the properties of many physical systems. For example, crystals, where the atoms are confined in the crystal sites with a well-defined spatial order. Or magnets, where the elementary magnets oscillate in phase and are mainly oriented in a given direction; the resulting magnetization characterizes the system of microscopic components as a whole. Without mentioning of course quantum optics and elementary particle physics.

In general, all systems that present an ordering in space or time (e.g., oscillating in phase) are regulated by microscopic dynamics characterized by coherence.

In the examples cited above, the concept of coherence is associated with the transition from the level of elementary components (microscopic level) to the level of the behavior of the system as a whole (macroscopic level). This transition from the microscopic to the macroscopic (or mesoscopic) scale is a very important and distinctive aspect of the mathematical formalism describing the phenomenon of coherence. It gives a quantitatively well-defined meaning to the notion of the emergence of a macroscopic property out of a microscopic dynamic process so that the macroscopic system possesses physical properties that are not found at the microscopic level [37, 38]. The behavior and the physical quantities that characterize the system as a whole are thus the results of the microscopic dynamics of the elementary components. Stiffness, for example, is a property of the crystal, not of its atomic or molecular components. The latter are confined to the crystal sites and cannot move freely, as an atom not belonging to a crystal would do; that is, they lose some of their degrees of freedom. However, characterization of the system at a macroscopic level (the stiffness of the crystal, its electrical conductivity, etc.) emerges dynamically from such freezing of microscopic degrees of freedom. The order, on the other hand, whether spatial or temporal, is itself a *collective* characteristic of a set of elementary components (it makes no sense to speak of order in the case of a few elementary components). It is therefore in this sense that we speak of *macroscopic quantum systems*. Crystals and magnets are examples of macroscopic quantum systems.

It should also be emphasized that, contrary to what is sometimes erroneously stated, in many systems the dynamic regime of coherence persists over a wide range of temperatures, from thousands of degrees centigrade to quite low temperatures, below zero centigrade. For example, diamond melts at the critical temperature T_C of 3545 °C, while sodium chloride crystals, the familiar kitchen salt, melt at 804 °C; in iron, the coherence between the elementary magnets which manifests itself in the magnetized state is lost at 770 °C, while in the cobalt, this occurs at 1075 °C (the critical transition temperature from the ferromagnetic to the non-magnetic phase is called the Curie temperature). In superconductors, on the other hand, the critical temperatures are very low, not higher than about −252 °C for some niobium compounds, and about −153°C for some superconductors discovered in the second half of the 1980s, such as certain copper oxides containing bismuth. The critical temperatures for the coherence phenomenon can therefore be very low or very high (compared to the ambient temperature), depending on specific conditions and dynamic properties characteristic of the system considered.

One further remark is that the coherence phenomenon preserves the macroscopic state from perturbations coming from quantum fluctuations. The latter are unavoidable at the level of the quantum dynamics of the elementary components of the system. However, in (Glauber) coherent states, we have $\langle \Delta N \rangle / \langle N \rangle = 1/|\alpha|$, where $\langle N \rangle$ denotes the number of elementary components in the coherent state, $\langle \Delta N \rangle$ the fluctuation of $\langle N \rangle$ and $|\alpha|$ is a measure of the coherence. We find that, for high $|\alpha|$, $\langle \Delta N \rangle$ is negligible compared with $\langle N \rangle$. This shows that coherence plays a crucial role in macroscopic stability against quantum fluctuations. The need to use fields, in particular quantum fields, comes from the fact that $\langle N \rangle$ is a large number for coherent states, indeed $\langle N \rangle = |\alpha|^2$, with high $|\alpha|$.

We thus see that the observed long lifetime of ordered systems, such as crystals, magnets, superconductors, etc. ("diamonds are forever" and kitchen salt "does not expire", that is, it can be kept for years, even in outdoor storage, and it is found in salt mines) is a result of the coherent dynamics of quantum fields (this is one of the major differences between QFT and quantum mechanics (QM) where the decoherence phenomenon occurs).

The degrees of freedom of the elementary components characterize the dynamics that regulate their spatial distribution and their evolution over time, and are in general closely associated with the symmetry properties of the dynamics [39]. For example, the possibility for an atom to be placed at any point in space without inducing observable variations in the system is described as space translational symmetry. Thanks to this symmetry, a set of atoms can assume different spatial configurations equivalent to each other from the point of view of observations, therefore physically equivalent, and in this respect indistinguishable from each other. In the example of the crystal, however, the atom is no longer free to be in "any" space position, but bound to sit in a specific crystal site. We then say that there is spontaneous breakdown of the symmetry (SBS). Here, spontaneous means that the state of the crystal is dynamically selected and generated among the possible accessible states.

In summary, the crystalline order results from the breaking of space translational symmetry; *order is lack of symmetry*. The different crystalline structures that are observed (cubic, rhombohedral, etc.) correspond to different ways of breaking the symmetry in the various spatial directions.

The conclusion is that, while symmetry describes the indistinguishability between states of the system linked by a symmetry transformation, order, i.e., the breaking of symmetry, allows one to distinguish between one state and another: the possibility of distinguishing, diversity, individuality of the state emerges from the establishment of order.

Much of the physics developed since the second half of the last century is based on the mechanism of symmetry breaking and the consequent formation of ordered structures, and this is linked to the notion of coherence. Quantum field theory (not QM) provides the mathematical formalism necessary for the study of spontaneously broken symmetry theories [37, 39, 40].

The Goldstone theorem in QFT states that SBS implies the existence of long-range correlations among the system elementary constituents. The quanta of these correlation waves are called Nambu–Goldstone (NG) bosons or quanta [41].

Boson particles can occupy the same physical state in any number (unlike fermion particles, where no more than one can occupy a given state, according to the Pauli exclusion principle). When many bosons sit in the same state, one says that they are *condensed* in that state. If they are massless, as NG bosons are, at their lowest (zero) momentum, they do not supply energy to that state, which can therefore be the least energy state of the system (also called the vacuum). If the condensed bosons are in phase, i.e., the long-range correlation waves of which they are the quanta are in phase, as happens in ordered states, the ground state is a coherent condensed state.

The ground state (or vacuum) of the system is then characterized by the non-zero expectation value of a quantity, characteristic of the symmetry which has been broken, called the order parameter since it is a measure of the ordering induced by the long-range correlations. In the crystal example, it is the density, in the ferromagnets the magnetization. The order parameter is a classical field of quantum origin, meaning that it is independent of quantum fluctuations. It is indeed a measure of the coherence of the system ground state generated by the Bose–Einstein *condensation* of NG bosons [37, 39–41].

The order parameter may assume different values in a given range and it depends on the temperature. Above a critical temperature T_C, it vanishes and the *phase transition* to the symmetric state is obtained with loss of the ordered structure (symmetry restoration). See above for examples of values of T_C (in diamonds, magnets, superconductors).

QFT thus allows the description of different phases in which the system may be found. These different phases present physically different types of behavior depending on the different values of the order parameter and are described by physically different spaces of states of the system, i.e., unitarily inequivalent representations of the canonical commutation relations (CCR).

In fact, infinitely many unitarily inequivalent representations of the CCR exist in QFT. They do not exist in QM due to the von Neumann theorem, which states that, for systems with a finite number of degrees of freedom, all the representations of the CCR are unitarily (and therefore physically) equivalent. Fields by definition describe systems with an infinite number of degrees of freedom, so the von Neumann theorem does not hold for them [37, 39, 40]. Systems that may have different physical phases need therefore to be described by QFT, which may account for the multiplicity of their phases and the transitions among them, not by QM.

4 The Many-Body Model of the Brain

Lashley's dilemma and the problems arising in the study of the brain, mentioned in Sect. 2, have their origin in the huge number of brain constituents at cellular and subcellular levels and in the great complexity of their organization and dynamics. The stability of the functional activity of the brain is, on the other hand, essential in any of our activities, and even for our survival in the world. How can it arise out of the myriads of brain constituents? There are of the order of 10^{11} neurons with 10^{15} synapses, each of them able to fire about 10 pulses per second, implying around 10^{16} synaptic operations per second, without mentioning glia cells and the fact that all this happens in a bath of 90% more numerous water molecules. Each of these molecules carries an oscillating electric dipole momentum subject to unavoidable quantum fluctuations.

However, the total activity of the brain requires an energy consumption per second of the order of only 25 W. This is ridiculously small compared with the power necessary for the simulation of quite elementary tasks by one of the gigantic American or European Brain Projects, which is of the order of 1.5 MW.

As already mentioned above in quoting Ricciardi and Umezawa, one should also explain how it happens that the brain's functional efficiency is not affected by the malfunctioning or even the loss of single neurons. Metabolic activity induces chemical transformations and replacements of biomolecules in intervals of time of the order of a couple of weeks. It is then hard to explain the long and medium lifetime of memories in terms of localized arrangements of biomolecules, due to their changes and renewal in such a turn-over process.

Schrödinger observed that the "enigmatic biological stability" [24, p. 47] of living matter (but, as already said, his observation may apply to the brain, too) cannot be explained in terms of "regularities only in the average" [24, p. 78] originating from the "statistical mechanisms". According to him, this would be the "classical physicist's expectation" that "far from being trivial, is wrong" [24, p. 19].

Starting from these remarks, in 1967 Ricciardi and Umezawa observed [23]:

> [...] it seems that very few concrete results have been obtained, in the sense that the question of *how the brain works out the information received from the outside, and what is the logic on which the operations performed by the brain are based* is still far from receiving a satisfactory solution. [...] One possibility then arises naturally: since one usually ignores the mechanism according to which the brain performs intelligent operations, [...] one could try to give a more general description of the brain dynamics; [...] from a phenomenological point of view it is strongly suggestive of a quantum model. In other terms, one can try to look for specific dynamical mechanisms (already known in the physics of many degrees of freedom) which can satisfy the essential requirements of the observed functioning of the brain.

In the many-body (quantum) model of the brain formulated by Ricciardi and Umezawa (RU), they assume that the external stimulus perceived by the brain is responsible for breaking the symmetry. The density of the NG correlation quanta generated by this breaking process (see Sect. 3) is assumed to be an index or distinctive code of the memory associated with the external stimulus that induced the symmetry breaking.

The same happens in the dissipative quantum model which will be discussed below. In both models the symmetry which is broken by external stimuli has been identified with the rotational (spherical) symmetry of the molecular electric dipoles [8, 27–30, 42].

It should be emphasized that, in the RU model (and in the dissipative model), neurons and other cellular units are classical systems. The quantum variables are the vibrational modes of the electric dipoles of the aqueous matrix and of the other biomolecules present. The long-range NG correlations among them promote and sustain the assembly and disassembly of oscillating domains of neuronal populations.

It is important to note that the stimulus does not affect the internal dynamics of the brain. It only induces the spontaneous breakdown of the rotational symmetry of the dipoles of water molecules. The internal dynamics then proceeds on the basis of the physical and chemical properties of the brain, independently of the stimulus. This aspect can have a clear and direct verification in the laboratory and its description constitutes a distinctive merit of the quantum model of the brain. It also accounts for the fact that an external stimulus, even dissimilar from the one originally inducing the memorization process, can stimulate the recall of the previously recorded memory [23, 43, 44]. This explains in dynamic terms the commonly experienced phenomenon of recalling a memory in perceptual conditions that are also very different from those in which it was first memorized. Here we are referring to normal or "weak" stimuli, not of a highly stressful type, such as a shock (or also an electro-shock) able to enslave the functionality of the brain. Although the stimulus can be quite weak, it does need to be "in phase" with the brain dynamics to induce SBS.

In the RU model and its subsequent developments [45, 46], the recording of a memory induced by a stimulus was canceled by that of another memory induced by a subsequent stimulus. This memory overprinting minimized the memory capacity of the model. For reasons of simplicity, the model did not consider the fact that the brain is a system in continuous, unavoidable interaction with the environment, intrinsically open to it. "Closing" the brain means damaging its functionality, as can be observed in experiments forcing a subject into isolation. The RU quantum brain model was therefore modified to include the "openness" of its dynamics. This led to the formulation of the dissipative quantum model of the brain [8, 9].

5 The Dissipative Quantum Model of the Brain

Any attempt to describe the brain cannot ignore the continuous and reciprocal energy exchange between the brain and the environment. Knowledge of the biomolecular and cellular details is fundamental but clearly insufficient for the description of brain activity. This alone cannot take into account the property of the system of being an open system.

As we have already seen, the detailed study of the elementary components is necessary, but not sufficient. It must be supplemented by knowledge of the dynamics that governs the set of elementary components. In the case of the brain, it is a dissipative form of dynamics. This leads us to the formulation of the dissipative quantum model of the brain [8, 9].

The physical need to consider the brain and, at the same time, its environment translates into the mathematical need to "double the system" [47]: we have the brain system and the environment system. The latter, which cannot be eliminated, can be schematized as the reservoir from which everything the brain absorbs comes from, and into which everything that the brain releases is poured. The overall brain–environment system is a closed one for which the energy flow at the brain–environment boundary is perfectly balanced.

From the standpoint of the balancing of flows, the environment is therefore described in the same way as the brain is described, provided that the "flow in" is changed into "flow out", and vice versa, which is obtained algebraically by changing the sign of the time variable: the environment is, therefore, a "time-reversed" copy of the brain.

Obviously, the interaction of the brain with the environment is very complex and requires the knowledge and detection of a huge number of parameters. However, if we limit ourselves to considering only the balance of the energy flows, the description of the environment as a time-reversed copy of the brain is mathematically correct and sufficient. In this description, the environment is therefore effectively a "copy", the *Double* of the brain.

The interplay between linearity and nonlinearity plays an interesting role in the dissipative model. Phase transitions between different representations (phases) occur in a nonlinear dynamical regime (criticality). SBS implies dynamical nonlinearity through which boson condensation and coherent states are formed. Linearity holds within each phase.

Such an interplay between linearity and nonlinearity is consistent with observations showing the coexistence of wave modes and pulse modes [48–50]. Pulse activity may be observed in experiments based on linear response. On the other hand, their synchronized AM patterns exhibit log–log power density versus frequency distributions, i.e., scale-free (self-similar fractal) dynamics requiring coherence consequent to nonlinear dynamics [49, 50]. Self-similar fractal properties are indeed isomorphic to coherent states [51–53], which is consistent with the underlying coherent many-body dynamics of the dissipative model [18, 54–57].

In the brain's dynamical evolution (in its "functioning"), there are variations in the flows exchanged with its environment, and therefore the state of the brain must be continuously updated, but so also must be the description of the state of the environment to balance the energy flows. In the memory states, which are two-mode coherent states, the brain modes are permanently entangled with the doubled modes (the environment modes) [43, 44, 58]. There is therefore a continuous "reciprocal updating", a process of reciprocal back-reaction, of "dialogue" between the brain and its Double, never a monologue, never resolvable. Sometimes in the conflict between the self and the Double, the dynamics of knowing, understanding, feeling, and living develop. The reciprocal influences of each on the other require a continuous updating of their relationship. Each of them is exposed to the gaze of the other [59].

It should be observed that the entanglement relation is implied by the in-phase correlation between the modes, which does not require exchange of a messenger or information and can therefore be established instantaneously without violation of the relativity postulates. Correlation is not therefore interaction, which would require the exchange of a messenger whose speed could not then be greater than the speed of light.

Returning to the dialog between the self and the Double, it is in this 'entre-deux' that the act of consciousness has its origin [8, 9]; it summarizes in itself the experience accumulated in the past, but is made up only of the present [9, 31, 59]. In this perspective, the brain appears to be "extended", in its own functionality, beyond the limits of its anatomical configuration. Consciousness expands into the environment in which the brain is immersed.

It is essential to stress that the relationship with the Double is a dynamical relationship, not one of narcissistic mirroring. In the dissipative model, there is nothing of such a mirroring. As Desideri observes [60], referring to certain discussions on mirror neurons [61, 62], mirroring is static and is not an opportunity for learning because the action observed and the action performed are structurally equivalent. What is observed in the laboratory [22, 63], and belongs to our common experience, is the property of the brain to accumulate experience and build knowledge, that is, to learn how to have "maximum grip" on the world. For this purpose, a copy, a simple mirroring is not enough; a creative operation is needed, a mimesis, in the sense of Aristotle's Poetics, which, as Desideri stresses, concerns the possible and not what simply happens. We need the amount of imaginative indeterminacy that allows learning and also a variation of the observed action model [60]. It is remarkable that the dissipative model allows such degrees of freedom and that the learning process arising from the dialogue with the Double is formally linked to minimizing the free energy of the system.

I observe that balancing the incoming and outgoing energy flows is equivalent to setting their difference to zero. This characterizes the state of equilibrium of the overall brain–environment system. However, setting the difference between two quantities to zero leaves them totally indeterminate. The balancing operation, therefore, allows an infinite series of pairs of states of the brain and the environment, respectively, for which such a difference is zero. Each of these brain states (and the corresponding environment states in each of the pairs) corresponds to a different value of the density of the condensed quanta. Each of these densities can be considered to be the index or code for a memory. It can be shown that states with different densities are orthogonal (unitarily inequivalent, see Sect. 3) to each other, therefore without mutual interference. Memories are thus protected form reciprocal "confusion". We see that the unitary inequivalence of the QFT representations thus plays a crucial role in the memory recording process. Moreover, their being infinite in number guarantees a large memory capacity. The result is that, thanks to dissipation, we may have infinitely many non-interfering memory states. Dissipation solves the memory capacity problem. The huge memory capacity is a consequence of the fact that the brain is a dissipative, open system [8].

6 Chaotic Trajectories in the Landscape of Attractors. Errare e Pensare

From what was said above, we see that the acquisition of a "new" memory corresponds to the use of one of the infinitely many unitarily inequivalent fundamental states (vacua) to which the brain–environment system has access. We can therefore describe the set of memory states (or "memory space") as the set of such coherent fundamental states, each one labeled by the code of a specific memory. These are states of minimum energy since, in them, the difference between the incoming and outgoing energy flows is zero, as mentioned above. Moreover, they are also states of minimum free energy. They are thus states towards which the system "tends" in its evolution, as towards "attractors". The set of memory states, therefore, depicts a "landscape of attractors".

It is remarkable that the strict mathematical unitary inequivalence among the representations is smoothed out in realistic situations due to defects, impurities, and surface effects. Such an "imperfection" is most welcome since it allows the evolution of memories in time and thus the possibility of "forgetting", and memories with different lifetimes, thus short-term memories and long-term memories. Moreover, it also allows transitions, or paths, trajectories through memories (through memory states), so that correlations may arise among the attractors as the brain goes from attractor to attractor; indeed along trajectories in the landscape of attractors, dwelling more or less for a long time in each of them, never, in normal (health) conditions, being trapped there. Along each trajectory, the free energy of the system is minimal. In the dissipative model, free energy and its minimization play a crucial role. This actually controls the density of the condensate labeling each memory state.

It needs to be stressed that the acquisition of a new memory involves not only the addition of a new attractor to the landscape of attractors, but the reorganization of the entire landscape, and therefore its complete updating in the light of the new acquisition.

It is in this process that the contextualization of the new acquisition and the emergence of its *meaning* consists, which never belongs to the perceptual stimulus (to the input). It belongs to the context of the redesigned landscape of attractors, always new as a whole. The meaning content of the correlations in the space of the attractors is therefore never definitive. Meanings can always be updated, better understood, or completely changed. They are always under test. Thus there emerges a dimension of novelty, surprise, even astonishment associated with suddenly seeing something unexpected [59, 64]. In this different view [65], one must seek the genesis of the imagination, and its role in determining different trajectories in the space of attractors. We are a long way from a simple mirroring. The relationship of the brain with the world is a completely dynamic one.

The process of contextualization, in which the brain calls into question its entire experiential history, constitutes one of the most salient features of the quantum dissipative model of the brain. It faithfully describes the laboratory observations in which the subject examined, animal or man, reacts to the situations in which he finds himself undergoing the process of abstraction (or exemplification necessary for the construction of the new attractor, i.e., the balancing of flows) and of generalization (or creation of categories in establishing correlations in the landscape of attractors). In this way the flow of information exchanged in the relationship with the world becomes knowledge [31, 59, 66] and memory becomes memory of meanings, not memory of information.

Each act of recognition of the attractor landscape represents an act of intuitive knowledge, the recognition of a *collective mode*, not divisible into rational steps, thinkable but "non-computational", and not translatable into the logical framework of language [59, 67]. This feature of brain activity is perhaps consistent with von Neumann's statement [68]:

> We require exquisite numerical precision over many logical steps to achieve what brains accomplish in very few short steps.

The trajectories in the landscape of attractors, from memory to memory, can be shown to be classical trajectories, although they "connect" quantum states. Moreover, they are chaotic trajectories, that is, they are not periodic (a trajectory never intersects itself) and trajectories that have different initial conditions never intersect; rather they are (exponentially) divergent. It can be shown formally that the chaoticity of the trajectories originates from the quantum nature of the memory states [38, 43, 44].

The role of chaos described by the dissipative model is confirmed by laboratory observations. Freeman has stressed that [69]:

> The chaos is evident in the tendency of vast collections of neurons to shift abruptly and simultaneously from one

complex activity pattern to another in response to the smallest of inputs [...] This changeability is a prime characteristic of many chaotic systems [...] In fact, we propose it is the very property that makes perception possible. We also speculate that chaos underlies the ability of the brain to respond flexibly to the outside world and to generate novel activity patterns, including those that are experienced as fresh ideas.

It is indeed interesting to note that the chaotic characteristics of the trajectories in the landscape of the attractors favours a high perceptual resolution. In fact, minimal differences in the perception (such as can occur in the recognition of images, smells, flavors, etc.) are recognized in a short time due to the divergence of the (chaotic) trajectories. Divergent trajectories are in fact easily recognizable as different. Small differences in the initial conditions would generate non-diverging trajectories in the absence of chaoticity, and the recognition of such differences would be much more difficult.

In its temporal evolution, the brain thus appears to be a system that "lives" on many microscopic configurations described by the minimum energy states corresponding to different memories, passing from configuration to configuration (from memory to memory) in its paths in the landscape of attractors (criticality of the phase transition dynamics). Even a weak external perturbation (a weak stimulus) can induce transitions through these least-energy states. In this way, occasional (random) perturbations play an important role in complex brain activity. On the other hand, one demonstrate the connection between the doubling of the degrees of freedom mentioned above and the quantum noise of the fundamental states. Nonzero double modes may indeed allow quantum effects arising from the imaginary part of the action, which would not appear at the classical level [70].

As just observed, the role of chaos and noise predicted by the dissipative model is confirmed in laboratory observations with particular reference to the resting state of the brain, whose dynamics shows fractal self-similarity [13, 19, 51, 71].

In conditions of low degree of openness of the brain toward the environment, e.g., while dreaming, or under the effects of psychoactive substances, during meditation, or in other states of reduced sensory perception, the criticality of the dynamics is enhanced [6, 7, 72]. Chaotic trajectories through the memory space then depict visual brain experiences occurring under such conditions. Indeed, these experiences are often characterized by movie-like sequences of images, with abrupt shifts from one image pattern to another. The truthfulness and realism felt in these visual experiences can be discussed in terms of the algebra of the doubling of the degrees of freedom. In the low openness states of the brain, the self almost fails to perceive the Double as distinct and their almost complete matching introduces a sort of "truth evaluation function" out of which the truthfulness and the realism of the visual experience is confirmed by the immediate and univocal feedback [6, 7].

The strong influence on trajectories due to minimal changes in their initial conditions leads us to consider the role of "doubt" [38, 59, 73] in the dissipative model. The dialogue with the Double lives on the continuous restructuring of the landscape of attractors, and this in turn can induce, in a process of self-questioning and listening [73], weak perturbations in the initial conditions of the trajectories with the consequent manifestation of their divergence. In this process, brain activity can be induced to leave other paths or to escape entrapment ("fixations") in one or another attractor. Doubts can well be understood as wandering around the landscape of attractors caused by the uncertainties linked to its constant redrawing, induced by the seduction of new perspectives opened by the unfolding of a new trajectory, questioning certainties acquired in previous perceptive experiences. It is therefore this wandering (*errare*) in search of *the possible*, not satisfied with what simply happens [60], a characteristic trait of brain activity, of thinking (of *pensare*). This is why the brain is not a stupid star. It rather behaves as an "erratic device", a "mistake-making machine" [74].

7 AI and Mistake-Making Machines. Spartacus

A machine is by definition and by construction a device that performs a succession of temporally linked operations in a strictly predictive way, like in a chain of logical steps. A machine that fails to go through such a determined chain of steps is therefore a machine that does not work properly and must be repaired or replaced. Brain activity, on the contrary, as we have seen, proceeds by steps that do not necessarily belong to a uniquely determined chain of steps.

Perhaps we might think then of the brain as a "mistake-making machine" [74]. Our great privilege of being able to make mistakes has its roots in the fact of wandering along chaotic trajectories in the attractor landscape, out of which the unpredictability of the movements of consciousness emerges, their being unfaithful to any pre-established scheme, their inalienable subjectivity, and total autonomy. As mentioned above, in the dissipative model the origin of the chaoticity of the trajectories lies in the quantum nature of the dynamics [38, 43, 44].

In AI research, the problem for the construction of an "intelligent" machine might therefore be just the problem of constructing a device able to make mistakes, an "erratic device". It is not a machine "not properly functioning" or "out of order". Its main usefulness is not in its predictable behavior, but rather in the "novelties" appearing in its behavior. The error, or mistake needs, however, to have the character of being "exceptional" with respect to the normal or "correct" behavior of the device (by definition, its stupid or boring behavior, Aristotle's stupid star). The "novel" or "intelligent" solution to a given problem proposed by the erratic device does not belong to the list of known possible solutions to possible problems, included in the device's basic instructions. Those are the predictable solutions of, e.g., the AI automatic pilot of an airplane. Such an automatic pilot must indeed be replaced by a human pilot in the case of an unforeseen emergency, requiring a solution which is "not on the list" of the automatic pilot.

In the above remarks, the reference to mistakes is not in the sense of observer-related mistakes, but to mistakes arising intrinsically in the behavior of erratic devices, not with respect to the expectations of the observer [74].

For observer-related mistakes, it is known that the problem of "right and wrong", "true and false", resides largely in the choice of the model adopted by the observer. Within the adopted model, the theory of the errors helps in the evaluation of the mistakes, also considering the possible interferences of the observer with the measurement and the phenomenon under study. The observer-related mistakes, in their departure from the observer's expectations, may have the character of "deviation" from the correct behavior. One might even define a trivial mistake-making machine, namely a machine doing other than what is expected by a single observer, and a non-trivial mistake-making machine, doing other than what is expected by any observer [74].

For the non-observer-related mistakes, the unpredictability of the mistake implies that it "cannot be expected or unexpected in any given context" [74]. Thus it is neither a negation, nor a deviation. It is "gratuitous", not "derivable", thus indicating a "non-computational" activity of the device.

It is interesting to note how a certain conservative attitude, confusing novelty with deviation, experiences the novelty as an attack on one's own model (status quo) and not as an addition to this, which may result in growth and strengthening of the model itself. However, the other alternative is not excluded, namely that the pure conservative, even if he does not confuse novelty with deviation, is against any possible novelty to be on the safe side and not take the risk that the novelty may invalidate the pre-existing model in whole or in part.

Summing up, we see that an intrinsic erratic device, producing non-observer-related mistakes as described, "cannot be a Turing machine, namely an algorithm generating mistakes. Indeed, it is not possible to design an algorithm doing nothing but the expected result, even if such a result is defined to be wrong with respect to certain criteria" [74].

The question then arises whether it is possible to design non-observer-related erratic devices. Remarkably, since the non-observer-related mistake is a novelty, "an emergence" in the system behavior, such a question may also be related to the one of designing "emergence", considered as a possible error appearing in mistake-making processes [74].

The alternative to the intrinsic erratic device would just be a prosthesis useful to help us in some of our physical or behavioral deficiencies or to improve or enhance our limited abilities (an artificial arm, possibly controlled through links to our nerves, a large capacity of memories with a only very short time needed to sort one of them out, a computer, an automatic pilot, a mobile phone, referred to as smart with a subtle sense of humor, a robot, etc.), which is more or less what AI provides at present.

Perhaps the program of constructing a conscious artificial device goes through the construction of the intrinsic erratic device. If it is ever going to be possible to build a device endowed with consciousness, it must possess all the best properties that characterize the human being: the unpredictability of his behavior, his ability to learn, but also his infidelity, his inevitable involvement with the world, and his inalienable freedom. And *he/she* must be called *Spartacus* [59, 74].

8 Concluding Remarks

Summarizing, we have seen that SBS implies the dynamical generation of long-range correlation waves among the elementary components of the system and that the associated quanta, the NG boson modes, are massless. Their coherent condensation in the system's ground state is responsible for the ordering observable there.

Symmetry corresponds to an invariance of the observable properties of the system; the system states before and after the symmetry transformation are "indistinguishable". Order, which results from a breakdown of symmetry, corresponds instead to the possibility of distinguishing the state of the system before and after the symmetry transformation; "diversity" thus arises. The scenario arising through SBS in QFT is one of great richness of *forms*. The dynamical processes leading to them, which we may refer to as *morphogenetic* processes, actually describe the dynamical generation of many different observable manifestations of the same basic dynamic equations, a "proliferation of differences" in the world around us. It is the richness of diversity, *the praise of Babel.*

Dissipative systems are not closed in on themselves. They exchange energy, momenta, mass, etc., with their environment. Then in each of them, the time translational invariance is broken, implying that the origin of the time axis cannot be freely translated, whence time is no longer a dummy variable [56, 75].

Dissipative systems are aging systems. History has its origin.

The many-body model of the brain and the dissipative quantum model of the brain were born by applying the QFT formalism of SBS to answer some of the open questions in neuroscience, as described in the sections above.

The doubling of the degrees of freedom required by the mathematical formalism for dissipative systems has led to the introduction of a "mirror in time" image of the self, or Double.

In the dialogue with the Double, knowledge is built on the basis of the experience accumulated in past perceptions. A perspective, or vision of the world, arises from this. The intentionality that determines our doing finds its root in updating a never definitive balance with the world around us, generating meanings and meaningful relationships with it.

It may happen that "the perfect exchange between inside and outside" is realized, a "favorable connection" between the self and the object. According to Desideri, this is the aesthetic experience [65]. The aesthetic one is therefore not a particular experience, nor just any experience, but [65] "it presents itself as a dimension that permeates the entire field of our experience (and the perceptual texture that configures the 'landscape')".

Recognizing such an experience determines the aesthetic judgment that "always involves the first person" [65]. The result is that of "taking a new look at the world" [65], which is not alien, but rather, a competitor with the cognitive dimension [31, 76]. The divergence of the trajectory in the landscape of the attractors in fact guarantees that the aesthetic experience is always new, and subversive compared to the consolidation of already explored landscapes. The orientation it expresses "is always awaiting renewal" [65] because the balancing of flows, of which it is an expression, is a dynamical one, never definitive [31]. The emotional response to the aesthetic experience thus possesses a performative value in the intentional arc [65]. The aesthetic experience is therefore a characteristic feature of brain dynamics [77].

Chaotic classical trajectories going through memory states characterize brain dynamics, offering the brain the ability to provide unpredictable behavior and answers to perceptual experiences. This leads us to depict the brain as an intrinsically erratic device, able to proceed by steps that are not linked by a strict, univocally determined succession. As mentioned in Sect. 6 and observed by Freeman [69], chaos does indeed play a relevant role in brain activity.

These properties of the brain's functional activity seem difficult to model within the framework of an AI research program. The same motivation for the project of an intrinsically erratic AI device is difficult to justify since such a device is in some sense just the opposite of the obedient, loyal machine pursued in actual AI projects. The perfect robot is required to be faithful and unfailing in pursuing the social, industrial, or military tasks justifying the financial efforts supporting its construction. Moreover, it needs to have a relatively short lifetime so as not to saturate the market (a key requirement already applied to smartphones, automobiles, TV sets, dishwashers, etc.). Once it has been constructed (and sold), its commercial value must tend rapidly to zero (and this is indeed the case). These realistic elements of "fragility" inherent in AI devices make it difficult for them to be more than prosthetics for some of our own physical deficiencies or disabilities, as commented in Sect. 7. Unfortunately, AI projects today are still limited to the design of "stupid stars".

I would like to close with an observation on the brain–environment frontier [78]. When the environment is made up of others, the question is: where do "I" end and where do "the others" begin? The question becomes even more radical when it comes from groups of people who feel they have a "strong identity": where do "we" end and where do "they" begin?

Can we imagine the world without others? One possible answer is: "No. We are all together, we ourselves are the others". Or the opposite answer: "Not them, just us. We come first, they are different from us". But the latter is not compatible with the openness of brain dynamics. The closure it proposes is equivalent to suicide. The others are part of our Double, too, namely of ourselves. "Their elimination" would be a self-elimination. We belong to each other; all the richness of imagination and creativity of the dialogue with the Double enters in our mutual relationship. The brain has an inherent social dimension.

Finally, it is perhaps interesting here to quote a passage from "Borges and I" [79] testifying to the broad imaginative horizon of the dialogue between the self and its Double:

> The other one, the one called Borges, is the one things happen to [...] It would be an exaggeration to say that

ours is a hostile relationship; I live, let myself go on living, so that Borges may contrive his literature, and this literature justifies me [...] Besides, I am destined to perish, definitively, and only some instant of myself can survive him [...] Spinoza knew that all things long to persist in their being; the stone eternally wants to be a stone and a tiger a tiger. I shall remain in Borges, not in myself (if it is true that I am someone) [...] Years ago I tried to free myself from him and went from the mythologies of the suburbs to the games with time and infinity, but those games belong to Borges now and I shall have to imagine other things. Thus my life is a flight and I lose everything and everything belongs to oblivion, or to him.

I do not know which of us has written this page.

References

1. Vitiello, G. (2016) Filling the gap between neuronal activity and macroscopic functional brain behavior. In R. Kozma, & W. J. Freeman (Eds.), *Cognitive phase transitions in the cerebral cortex—Enhancing the neuron doctrine by modeling neural fields* (pp. 239–249). Springer Int. Pub. Switzerland.
2. Pribram, K. H. (1971). *Languages of the brain*. Prentice-Hall, Engelwood Cliffs NJ.
3. Pribram, K. H. (1991). *Brain and perception*. Lawrence Erlbaum.
4. Pribram, K. H. (2013). *The form within: My point of view*. Prospecta Press.
5. Freeman, W. J. (1997). Nonlinear neurodynamics of intentionality. *Journal Mind Behaviour, 18*, 291–304.
6. Re, T., & Vitiello, G. (2020). Nonlinear dynamics and chaotic trajectories in brain–mind visual experiences during dreams, meditation and non-ordinary brain activity states. *OBM Neurobiology, 4*(2). https://doi.org/10.21926/obm.neurobiol.2002061.

7. Re, T., & Vitiello, G. (2020) On the brain–mind visual experiences. In *Proceedings IEEE—IJCNN, Glasgow*. https://doi.org/10.1109/IJCNN48605.2020.9207327.
8. Vitiello, G. (1995). Dissipation and memory capacity in the quantum brain model. *International Journal of Modern Physics B, 9*, 973–989.
9. Vitiello, G. (2001). *My double unveiled*. John Benjamins.
10. Lashley, K. S. (1942). The problem of cerebral organization in vision. VII, Visual mechanisms *Biological Symposia* (pp. 301–322). Jaques Cattell Press.
11. Freeman, W. J. (2001). *How brains make up their minds*. Columbia University Press.
12. Freeman, W. J. (2005). Origin, structure, and role of background EEG activity. Part 3. Neural frame classification. *Clinical Neurophysiology, 116*, 1117–1129.
13. Freeman, W. J. (2006). Origin, structure, and role of background EEG activity. Part 4. Neural frame simulation. *Clinical Neurophysiology, 117*, 572–589.
14. Freeman W. J., Gaàl G., & Jornten R.: A neurobiological theory of meaning in perception. Part 3. Multiple cortical areas synchronize without loss of local autonomy. *International Journal of Bifurcation and Chaos in Applied Sciences and Engineering, 13*, 2845–2856.
15. Freeman, W. J., & Rogers, L. J. (2003). A neurobiological theory of meaning in perception. Part 5. Multicortical patterns of phase modulation in gamma EEG. *International Journal of Bifurcation and Chaos in Applied Sciences and Engineering, 13*, 2867–2887.
16. Freeman, W. J. (2004). Origin, structure, and role of background EEG activity. *Part 1. Phase Clinical Neurophysiology, 115*, 2077–2088.
17. Freeman, W. J. (2004). Origin, structure, and role of background EEG activity. *Part 2. Amplitude Clinical Neurophysiology, 115*, 2089–2107.
18. Freeman, W. J., & Vitiello, G. (2006). Nonlinear brain dynamics as macroscopic manifestation of underlying many-body dynamics. *Physics of Life Reviews, 3*, 93–117. q-bio.OT/0511037

19. Freeman, W. J., & Vitiello, G. (2008). Dissipation, spontaneous breakdown of symmetry and brain dynamics. *Journal Physics A: Mathematical Theory, 41*, 304042. q-bio. NC/0701053.

20. Bassett, D. S., Meyer-Lindenberg, A., Achard, S., Duke, T., & Bullmore, E. (2006). Adaptive reconfiguration of fractal small-world human brain functional network. *PNAS, 103*, 19518–19523.

21. Fingelkurts, A. A., & Fingelkurts, A. A. (2004). Making complexity simpler: Multivariability and metastability in the brain. *International Journal of Neuroscience, 114*, 843–862.

22. Freeman, W. J. (1975/2004). *Mass action in the nervous system*. Academic Press, New York.

23. Ricciardi, L. M., & Umezawa, H. (2004). Brain and physics of many-body problems. *Kybernetik, 4*, 44–48 (1967). Reprint in G. G. Globus, K. H. Pribram, & G. Vitiello (Eds.) *Brain and Being* (pp. 255–266). John Benjamins, Amsterdam.

24. Schrödinger, E. (1944). *What is life? [1967 reprint]*. Cambridge: University Press.

25. Fröhlich, H. (1968). Long range coherence and energy storage in biological systems. *International Journal of Quantum Chemistry, 2*, 641–649.

26. Hyland, G. H. (2015). *Herbert Fröhlich. A physicist ahead of his time*. AG Switzerland: Springer International Publishing.

27. Del Giudice, E., Doglia, S., Milani, M., & Vitiello, G. (1983). Spontaneous symmetry breakdown and boson condensation in biology. *Physics Letters, 95A*, 508–510.

28. Del Giudice, E. Doglia, S. Milani, M., & Vitiello, G. (1985). A quantum field theoretical approach to the collective behavior of biological systems. *Nuclear Physics, B251* (FS 13), 375–400 (1985)

29. Del Giudice, E. Doglia, S. Milani, M., & Vitiello, G. (1986). Electromagnetic field and spontaneous symmetry breakdown in biological matter. *Nuclear Physics, B275* (FS 17), 185–199.

30. Del Giudice, E., Preparata, G., & Vitiello, G. (1988). Water as a free electron laser. *Physical Review Letters, 61*, 1085–1088.
31. Vitiello, G. (2008). Essere nel mondo: Io e il mio Doppio. Atque 5 Nuova Serie, 155–176.
32. Umezawa, H. (1995). Development in concepts in quantum field theory in half century. *Mathamatical Japonica, 41*(1), 109–124.
33. Vitiello, G. (2011). Hiroomi Umezawa and quantum field theory. *NeuroQuantology, 9*(3), 402–412.
34. Atmanspacher, H. (2018) Walter freeman. I did it my way. *Journal Consciousness Studies, 25*(1–2), 39–44.
35. Atmanspacher, H. (2020). Quantum approaches to consciousness. *Stanford Encyclopedia of Philosophy*. http://plato.stanford.edu/entries/qtconsciousness/.
36. Vitiello, G. (1998). Structure and function. An open letter to Patricia Churchland. In S. R. Hameroff, A. W. Kaszniak, & A. C. Scott (Eds.), *Toward a science of consciousness II. The second Tucson Discussions and debates* (pp. 231—236). Cambridge: MIT Press.
37. Umezawa, H. (1993). *Advanced field theory: Micro, macro and thermal concepts*. American Institute of Physics.
38. Vitiello, G. (2004). Classical chaotic trajectories in quantum field theory. *International Journal of Modern Physics B, 18*, 785–792.
39. Itzykson, C., & Zuber, J. (1980). *Quantum field theory*. McGraw-Hill.
40. Blasone, M., Jizba, P., & Vitiello, G. (2011). *Quantum field theory and its macroscopic manifestations*. Imperial College Press.
41. Goldstone, J., Salam, A., & Weinberg, S. (1962). Broken symmetries. *Physical Review, 127*, 965–970.
42. Jibu, M., & Yasue, K. (1995). *Quantum brain dynamics and consciousness*. Amsterdam, The Netherlands: John Benjamins.
43. Pessa, E., & Vitiello, G. (2003). Quantum noise, entanglement and chaos in the quantum field theory of mind/brain states. *Mind Matter, 1*, 59–79.

44. Pessa, E., & Vitiello, G. (2004). Quantum noise induced entanglement and chaos in the dissipative quantum model of brain. *International Journal of Modern Physics B, 18,* 841–858.
45. Stuart, C. I. J., Takahashi, Y., & Umezawa, H. (1978). On the stability and non-local properties of memory. *Journal of Theoretical Biology, 71,* 605–618.
46. Stuart, C. I. J., Takahashi, Y., & Umezawa, H. (1979). Mixed system brain dynamics: Neural memory as a macroscopic ordered state. *Foundations of Physics, 9,* 301–327.
47. Celeghini, E., Rasetti, M., & Vitiello, G. (1992). Quantum dissipation. *Annalen der Physik, 215,* 156.
48. Capolupo, A., Freeman W. J., & Vitiello, G. (2013). Dissipation of dark energy by cortex in knowledge retrieval. *Physics of Life Reviews, 10,* 8594.
49. Freeman, W. J., Capolupo, A., Kozma, R., Olivares del Campo, A, & Vitiello, G. (2015). Bessel functions in mass action modeling of memories and remembrances. *Physics Letters, A379,* 2198–2208.
50. Capolupo A, Kozma R, Olivares del Campo A, & Vitiello G. (2017). Bessel-like functional distributions in brain average evoked potentials. *Journal of Integrative Neuroscience, 16,* S85–S98.
51. Vitiello, G. (2009). Coherent states, fractals and brain waves. *New Mathematics and Natural Computing, 5,* 245–264.
52. Vitiello, G. (2012). Fractals, coherent states and self-similarity induced noncommutative geometry. *Physics Letters A, 376,* 2527–2532.
53. Vitiello, G. (2014). On the isomorphism between dissipative systems, fractal self-similarity and electrodynamics. Toward an integrated vision of nature. *Systems, 2,* 203–216.
54. Vitiello, G. (2015) The use of many-body physics and thermodynamics to describe the dynamics of rhythmic generators in sensory cortices engaged in memory and learning. *Current Opinion in Neurobiology, 31,* 7–12.

55. Freeman W. J., & Vitiello, G. (2010). Vortices in brain waves. *International Journal of Modern Physics A, B 24*, 326995.
56. Freeman, W. J., & Vitiello, G. (2016). Matter and mind are entangled in two streams of images guiding behavior and informing the subject through awareness. *Mind and Matter, 14*, 7–24.
57. Vitiello, G. (2018). The brain and its mindful double. *Journal Consciousness Studies, 25*, 151–176.
58. Sabbadini, S. A., & Vitiello, G. (2019). Entanglement and phase-mediated correlations in quantum field theory. Application to brain–mind states. *Applied Science, 9*, 3203. https://doi.org/10.3390/app9153203.
59. Vitiello, G. (2004). The dissipative brain. In G. G. Globus, K. H. Pribram, & G. Vitiello (Eds.), *Brain and being* (pp. 255–266). John Benjamins.
60. Desideri, F. (2008). Del comprendere. A partire da Wittgenstein. *Atque 5 Nuova Serie*, 135–154.
61. Rizzolati, G., Fogassi, V., & Gallese, V. (2001). Neurophysiological mechanisms underlying the understanding and imitation of action. *Nature Neuroscience Reviews, 2*, 661–670.
62. Rizzolati, G., & Craighero, L. (2004). The mirror-neuron system. *Annual Review of Neuroscience, 27*, 169–192.
63. Freeman, W. J. (2005). NDN, volume transmission, and self-organization in brain dynamics. *Journal Integrative Neuroscience, 4*(4), 407–421.
64. Vitiello, G. (2006). Oggetto, percezione e astrazione in fisica. In F. Desideri & G. Matteucci (Eds.), *Dall'oggetto estetico all'oggetto artistico* (pp. 11–21). University Press.
65. Desideri, F. (2007). Il nodo percettivo e la meta-funzionalità dell'estetico. In F. Desideri & G. Matteucci (Eds.), *Estetiche della percezione* (pp. 13–24). University Press.
66. Atmanspacher, H., & Scheingraber, H. (1990). Pragmatic information and dynamical instabilities in a multimode continuous-wave dye laser. *Canadian Journal of Physics, 68*, 728–737.

67. Piattelli-Palmarini, M., & Vitiello, G. (2015). Linguistics and some aspects of its underlying dynamics. *Biolinguistics, 9*, 96–115.
68. Neumann von, J. (1958). *The Computer and the brain* (p. 63). New Haven: Yale University Press.
69. Freeman, W. J. (1991). The physiology of perception. *Scientific American, 264*, 78–87.
70. Srivastava, Y. N., Vitiello, G., & Widom, A. (1995). Quantum dissipation and quantum noise. *Annals of Physics, 238*, 200–207.
71. Freeman, W. J. (2000). *Neurodynamics*. An exploration of mesoscopic brain dynamics. Springer.
72. Alfinito, E., & Vitiello, G. (2000). Formation and lifetime of memory domains in the dissipative quantum model of brain. *International Journal of Modern Physics B, 14*, 853–868.
73. Desideri, F. (1998). L'ascolto della coscienza. Una ricerca filosofica. Milano: Feltrinelli.
74. Minati, G., & Vitiello, G. (2006). Mistake making machines. In G. Minati, E. Pessa, & M. Abram (Eds.), *Systemics of emergence: Research and development* (pp. 67–78). Springer.
75. Vitiello, G. (2016). ... and Kronos ate his sons .... In I. Licata & G. 't Hooft (Ed.), *Beyond peaceful coexistence. The emergence of space, time and quantum* (pp. 465–486). London: Imperial College Press.
76. Diodato, R. (1997). Vermeer, Góngora, Spinosa. L'estetica come scienza intuitiva. Milano: Bruno Mondatori.
77. Vitiello, G. (2015). The aesthetic experience as a characteristic feature of brain dynamics. *Aisthesis, 8*(1), 71–89. https://oajournals.fupress.net/index.php/aisthesis/article/view/859/85.
78. Vitiello, G. (2008). Campo dei fiori. In P. Poinsotte & M. Hack (Eds.) Scienza e Società (pp. 79–98). Aracne, Roma.
79. Borges J. L. (1960). Borges and I. In El hacedor, Biblioteca Borges, Alianza Editorial.

Hard Problem and Free Will: An Information-Theoretical Approach

Giacomo Mauro D'Ariano and Federico Faggin

We are such stuff as dreams are made on, and our little life is rounded with a sleep.

William Shakespeare

Abstract We explore definite theoretical assertions about consciousness, starting from a non-reductive psycho-informational solution of David Chalmers's *hard problem*, based on the hypothesis that a fundamental property of "information" is its experience by the supporting "system". The kind of information involved in consciousness needs

---

G. M. D'Ariano (✉)
Dipartimento Di Fisica, University of Pavia, via Bassi 6, 27100 Pavia, Italy
e-mail: dariano@unipv.it

F. Faggin
Federico and Elvia Faggin Foundation, San Francisco, USA

© The Author(s) 2022
R. Penrose et al., *Artificial Intelligence Versus Natural Intelligence*,
https://doi.org/10.1007/978-3-030-85480-5_5

to be quantum for multiple reasons, including its intrinsic privacy and its power of building up thoughts by entangling qualia states. As a result we reach a quantum-information-based panpsychism, with classical physics supervening on quantum physics, quantum physics supervening on quantum information, and quantum information supervening on consciousness.

We then argue that the internally experienced quantum state, since it corresponds to a definite experience–not to a random choice–must be pure, and we call it *ontic*. This should be distinguished from the state predictable from the outside (i.e., the state describing the knowledge of the experience from the point of view of an external observer), which we call *epistemic* and is generally mixed. Purity of the ontic state requires an evolution that is purity preserving, namely a so-called *atomic* quantum operation. The latter is generally probabilistic, and its particular outcome is interpreted as free will, which is unpredictable even in principle since quantum probability cannot be interpreted as lack of knowledge. We also see how the same purity of state and evolution allow one to solve the well-known *combination problem* of panpsychism.

Quantum state evolution accounts for a *short-term buffer of experience* and itself contains quantum-to-classical and classical-to-quantum information transfers. Long term memory, on the other hand, is classical, and needs memorization and recall processes that are quantum-to-classical and classical-to-quantum, respectively. Such processes can take advantage of multiple copies of the experienced state re-prepared with "attention", and therefore allowing for a better quality of classical storage.

Finally, we explore the possibility of experimental tests of our theory in the cognitive sciences, including the evaluation of the number of qubits involved, the existence of complementary observables, and violations of local-realism bounds.

In the appendices we succinctly illustrate the *operational probabilistic theory* (OPT) framework for possible post-quantum theories of consciousness, assessing the convenient black-box approach of the OPT, along with its methodological robustness in separating objective from theoretical elements, guaranteeing experimental control and falsifiability. We finally synthetically compare the mathematical postulates and theorems of the most relevant instances of OPTs–i.e. classical and quantum theories–to help the reader get a better understanding our theory of consciousness. The mathematical notation is provided in a handy table in the appendices.

1 A Quantum-Informational Panpsychism

In his book *The Character of Consciousness* [1] David Chalmers states what he calls the *hard problem of consciousness*, namely the issue of explaining our *experience*–sensorial, bodily, mental, and emotional, including any stream of thoughts. Chalmers contrasts the *hard problem* with the *easy problems* which, as in all sciences, can be tackled in terms of a mechanistic approach that is useless for the problem of experience. Indeed, in all sciences we always seek explanations in terms of *functioning*, a concept that is entirely independent from the notion of *experience*. Chalmers writes:

Why is the performance of these functions accompanied by experience?
Why doesn't all of this information processing go on "in the dark" free of any inner feel?
There is an explanation gap between the function and the experience.

An effective paradigm for grasping the conceptual gap between "experience" and "functioning" is that of the *zombie*, which is behaviourally indistinguishable from a conscious being, but nevertheless has no inner experience.

There are currently two main lines of response to the hard problem: (1) the *physicalist view*–with consciousness "emergent from a functioning", such as some biological property of life [2]; (2) the *panpsychist view*–with consciousness as a fundamental feature of the world that all entities have. What is proposed here is:

Panpsychism with consciousness as a fundamental feature of "information", and physics supervening on information.

The idea that physics is a manifestation of pure information processing has been strongly advocated by John Wheeler [3] and Richard Feynman [4, 5], along with several other authors, including David Finkelstein [6], who was particularly fond of this idea [7]. Only quite recently, however, has the new informational paradigm for physics been concretely established. This program achieved: (1) the derivation of quantum theory as an information theory [8–10], and (2) the derivation of free quantum field theory as emergent from the nontrivial quantum algorithm with denumerable systems with minimal algorithmic complexity [11, 12].[1] In addition to such methodological value, the new information-theoretic derivation of quantum field theory is particularly promising for establishing a theoretical framework for quantum gravity as emergent from the

---

[1] The literature on the informational derivation of free quantum field theory is extensive, and, although not up to date, we suggest the review [7] written by one of the authors in memory of David Finkelstein. The algorithmic paradigm has opened for the first time the possibility of avoiding physical primitives in the axioms of the physical theory, allowing a re-foundation of the whole of physics on logically firm ground [13].

quantum information processing, as also suggested by the role played by information in the holographic principle [14, 15]. To sum up, the physical world emerges from an underlying algorithm, and the kind of information that is processed beneath it is quantum.

The idea that quantum theory (QT) could be regarded as an information theory is a relatively recent one [16], and originated within the field of *quantum information* [17]. Meanwhile what we call "information theory" has largely evolved, from its origins as a communication theory [18], toward a general theory of "processing" of information, which had previously been the sole domain of computer science.

What do we mean by "information theory"?

Recently, both in physics and in computer science (which in the meantime connected with quantum information), the theoretical framework for all information theories emerged in the physics literature in terms of the notion of *operational probabilistic theory* (OPT) [9, 10, 19], an isomorphic framework that emerged within computer science in terms of *category theory* [20–25]. Indeed, the mathematical framework of an information theory is precisely that of the OPT, whichever information theory we consider–either classical, quantum, or "post-quantum". The main structure of OPT is reviewed in the Appendix.[2]

Among the information theories, classical theory (CT) plays a special role. In fact, besides being itself an OPT, CT enters the operational framework in terms of objective outcomes of the theory, which for causal OPTs (like QT and CT) can be used to conditioning the choice of a

---

[2] The reader who is not familiar with the notion of OPT can regard the OPT as the mathematical formalization of the rules for building quantum information circuits. For a general idea about OPT it is recommended to read the appendix. The reader is assumed to be familiar with elementary notions, such as *state* and *transformation* with finite dimensions.

subsequent transformation within a set. Clearly this also happens in the special case of QT. Thus, the occurrence of a given outcome can be regarded as a quantum-to-classical information exchange, whereas conditioning constitutes a classical-to-quantum information exchange. We conclude that we should regard the physical world as faithfully ruled by both quantum and classical theories together, with information transforming between the two types.

This theoretical description of reality should be contrasted with the usual view of reality as being quantum, creating a fallacy of misplaced concreteness. The most pragmatic point of view is to regard QT and CT together as the correct information theory to describe reality, without incurring any logical paradox. We will use this idea in the rest of the chapter. Due to the implicit role played by CT in any OPT, when we mention CT we intend to designate the special corresponding OPT.

We will call the present view, with consciousness as fundamental for information and physics supervening on quantum information, *quantum-information panpsychism*.

In place of QT, one may consider a *post-quantum* OPT– e.g. RQT (QT on real Hilbert spaces), FQT (fermionic QT), PRB (an OPT built on Popescu-Rohrlich boxes [26]), etc. [10], yet some of the features of the present consciousness theory can be translated into the other OPT, as long as the notion of "entanglement" survives in the considered OPT.

2   A Non-reductive Psycho-Informational Solution: General Principles

The fundamental nature of the solution to the hard problem proposed here has been suggested by David Chalmers as satisfying the following requirements [1]:

Chalm$_1$: *Consciousness as fundamental entity, not explained in terms of anything simpler ...*

Chalm$_2$: *a non reductive theory of experience will specify basic principles that tell us how basic experience depends on physical features of the world.*

Chalm$_3$: *These psychophysical principles will not interfere with physical laws (closure of physics). Rather they will be a supplement to physical theory.*

In an information-theoretic framework, in which physics supervenes on information, our principle will be *Psychoinformational* (see Chalm$_3$). The non-reductive (Chalm$_2$) Psychoinformational principle that is proposed here is the following:

P1: Psychoinformational principle: *Consciousness is the information system's experience of its own information state and processing.*

As we will soon see, it is crucial that the kind of information that is directly experienced be quantum.

Principle P1 may seem ad hoc, but the same happens with the introduction of any fundamental quantity (Chalm$_{1,2}$) in physics, e.g., the notions of *inertial mass*, *electric charge*, etc. Principle P1 asserts that *experience is a fundamental feature of information*, hence also of physics, which supervenes on it. P1 is not reductive (Chalm$_2$), and it does not affect physics (Chalm$_3$), since the kind of information involved in physics is quantum + classical. On the other hand, P1 supplements physics (Chalm$_3$), since the latter supervenes on information theory.

It is now natural to ask: *what are the systems?* Indeed, information is everywhere: light strikes objects and thereafter reaches our eyes, providing us with information about those objects: colour, position, shape, and so on; information is supported by a succession of systems: the light modes, followed by the retina, then the optical nerves, and finally the several bottom-up and top-down visual processes occurring in the brain. Though the final answer may have to come from neuroscience, molecular biology, and cognitive science experiments, we can use the present OPT approach to inspire crucial experiments. OPT has the power of being a *black-box* method that does not need the detailed "physical" specification of the systems, and this is a great advantage! Moreover, our method tackles the problem in terms of a purely in-principle reasoning, independent of the "hardware" supporting information, exactly as we do in information theory. Such *hardware independence* makes the approach particularly well suited to address a problem as fundamental as the problem of consciousness.

We now proceed with the second principle:

P2: Privacy principle: *experience is not shareable, even in principle.*

Principle P2 plays a special role in selecting which information theories are compatible with a theory of consciousness. Our experience is indeed not shareable: this is a fact. We hypothesize that that non shareability of experience holds *in principle*, not just as a technological limitation.[3] A crucial fact is that *information shareability* is equivalent to *information readability with no disturbance*.[4] Recently

---

[3] It may be possible to know which systems are involved in a particular experience, as considered in Ref. [27]. However, we could never know the experience itself, since it corresponds to non sharable quantum information.

[4] In fact, information shareability is equivalent to the possibility of making copies of it–technically *cloning information*. The possibility of cloning information, in turn, is equivalent to that of *reading information without disturbing it*. Indeed, if one could read information without disturbing it, one could read the information as many times as needed in order to acquire all its various complementary aspects, technically performing a *tomography* of the information. And once one knows all the information one can prepare copies of it at will. Viceversa, if one could clone quantum information, one could read the clones, keeping the original untouched.

it has been proved that the only theory where any information can be extracted without disturbance is classical information theory [28]. We conclude that P2 implies that a theory of consciousness needs nonclassical information theory, namely QT, or else a "post-quantum" OPT.

Here we will consider the best known instance of OPT, namely QT. As we will see, such a choice of theory turns out to be very powerful in accounting for all the main features of consciousness. We state this choice of theory as a principle:

P2': Quantumness of experience: *the information theory of consciousness is quantum theory.*

Note that classical theory always accompanies any OPT, so there will be exchanges and conversions of classical and quantum information.

We now introduce a third principle:

P3: Psycho-purity principle: *the state of the conscious system is pure.*

P3 may seem arbitrary if one misidentifies *experience* with the *knowledge of it*. The actual experience is *ontic* and *definite*. It is *ontic*, à la Descartes (*Cogito, ergo sum* ...). The existence of our experience is surely something that everybody would agree on, something we can be sure of. It is *definite*, in the sense that one has only a single experience at a time—not a probability distribution of experiences. The latter would describe the *knowledge of somebody else's experience*. The experience of a *blurred coin* is just as definite as the experience of a *sharp coin* since they are two different definite experiences. It is knowledge of the face of the coin that is definite, and this is only achieved if the image of the coin is sufficiently sharp, otherwise it would be represented in probabilistic terms, e.g. by the mixed

Ontic State
Experienced by the system

Epistemic State
Predicted by an outside observer

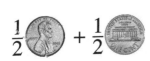

Fig. 1 Illustration of the notions of *ontic* and *epistemic* states for a system given by a classical bit, here represented by a coin, with states $0 = $ HEAD or $1 = $ TAIL. We call *ontic* the "actual" state of the system, which is pure and generally unknown, except as an unshareable experience (in the figure it is the coin state HEAD covered by the hand). We call *epistemic* the state that represents the knowledge about the system of an outside observer, e.g. the state of the unbiased coin corresponds to $\frac{1}{2}$ HEAD $+ \frac{1}{2}$ TAIL

state $\frac{1}{2}$ TAIL $+ \frac{1}{2}$ HEAD, based on a fair-coin hypothesis (see Fig. 1). We can thus understand that, by the same definition, the notion of mixture is an *epistemic* one, based on our prior knowledge about the coin having two possible states. In short:

1. *Experience* is described by an *ontic quantum state*, which is *pure*.
2. *Knowledge* is described by an *epistemic quantum state*, which is generally *mixed*.

We will see later how principle P3 is crucial for solving the *combination problem*.

We now state our fourth principle, which is theory independent:

P4: Qualia principle: *experience* is made of structured qualia.

Qualia (phenomenal qualitative properties) are the building blocks of conscious experience. Their existence has so far remained inexplicable within the traditional scientific framework. Their privacy and ineffability are notorious. Qualia are *structured*: some of them are more fundamental than others, e.g., spectrum colours, sound pitch, the five different kinds of taste buds (sweet, salty, sour, bitter, and umami), somatic sensations (pain, pleasure, etc.), basic emotions (sadness, happiness, etc.). Qualia are compositional, with internal structures that generally partially determine their qualitative nature. They are connected by different relations, forming numerous complex structures, such as thoughts and emotions. Since we regard consciousness as the direct fruition of a very structured kind of quantum information (we are the system supporting such final information), then, according to principle P1, qualia correspond to information states "felt" by the aware system, and according to principles P2 and P2', such states are quantum states of such systems.

S1: Qualia are described by ontic pure quantum states.

We will see in Sect. 3 how quantum entanglement can account for organizing nontrivial qualia. The *qualia space* would thus be the Hilbert space of the systems involved in the quale.

Classical information coming from the environment through the senses is ultimately converted into quantum information which is experienced as qualia. The inverse quantum-to-classical transformation is also crucial in converting structures within qualia into logical and geometric representations expressible with classical information to communicate free-will choices to classical structures. For such purposes, the interaction with the memory of past experiences may be essential, for, as we shall see, memory is classical.

Finally, we come to the problem of *free will*. First we must specify that free will, contrary to consciousness, produces public effects which are classical manifestations of quantum information. Since the manifestation is classical, the possible choices of the conscious agent are in principle *jointly perfectly discriminable*.[5]

We then state what we mean by *free*.

S2: Will is *free* if its unpredictability by an external observer cannot be interpreted in terms of lack of knowledge.[6]

In particular, statement S2 implies that the actual choice of the entity exercising its free will cannot in principle be predicted with certainty by any external observer. One can immediately recognise that if one regards the choice as a random variable, it cannot be a classical one, which can always be interpreted as lack of knowledge. On the other hand, it fits perfectly the case of quantum randomness, which cannot be interpreted as such.[7]

---

[5] In the usual classical information theory, the convex set of states is a simplex, and its extremal states are jointly discriminable.

[6] Here there is a clear distinction between the "willing agent" and the observers of it. The word *unpredictability* applies only to the observers.

[7] The impossibility of interpretation of quantum randomness as lack on knowledge fits quantum complementarity. Indeed there is no way of knowing both values of two complementary variables. One might argue that both values may exist anyway, even if we cannot know them, but such an argument disagrees with the violation of the CHSH bound (the most popular Bell-like bound), which is purely probabilistic, and is based on the assumption of the existence of a joint probability distribution for the values of complementary observables, an assumption that obviously is violated by quantum complementarity. Others argue (and this is the most popular argument) that one must use different local random variables depending on remote settings, which leads to the interpretation of the CHSH correlations in terms of *nonlocality*, but such an interpretation is artificial. The most natural argument is that *the measurement outcome is created by the measurement*.

3 Qualia: The Role of Entanglement

According to statement S1, qualia are described by *ontic* quantum states and, being *pure* states, we can represent them by normalised vectors $|\psi\rangle \in \mathcal{H}_A$ in the Hilbert space $\mathcal{H}_A$ of system A. For the sake of illustration we consider two simple cases of qualia: *direction* and *colour*. These examples should not be taken literally,[8] but only for the sake of illustrating of the concept that the linear combinations of qualia can give rise to new complex qualia.

We have said that qualia combine to make new qualia, and *thoughts* and *emotions* themselves are structured qualia. Superimposing two different kinds of qualia in an entangled way produces new kinds of qualia. Indeed, consider the superposition of a red up-arrow with a green down-arrow. This is not a yellow right-arrow as one might expect, since the latter corresponds to the independent superposition of direction and colour, as in the following equation:

$$\tfrac{1}{\sqrt{2}}(|\uparrow\rangle \otimes |\bullet\rangle + |\downarrow\rangle \otimes |\circ\rangle) \neq \tfrac{1}{2}(|\uparrow\rangle + |\downarrow\rangle) \otimes (|\bullet\rangle + |\circ\rangle) = |\rightarrow\rangle \otimes |\,\rangle \tag{1}$$

We would have instead

$$\tfrac{1}{\sqrt{2}}(|\uparrow\rangle \otimes |\bullet\rangle + |\downarrow\rangle \otimes |\circ\rangle) = \tfrac{1}{\sqrt{2}}(|\uparrow\rangle + |\downarrow\rangle) = |\bigstar\rangle, \tag{2}$$

where the ket with the blue star represents a completely new qualia. In a more general case, we have a state vector

---

[8] We emphasize that the above examples are only intended to illustrate the concepts. The example of color qualia in Fig. 2 would be a faithful one if the colours were monochromatic and the summation or subtraction were made with wave amplitude, not intensity, which is what actually happens. The case of direction qualia based on the Bloch sphere is literally correct. The directions are those of the state representations on the Block sphere.

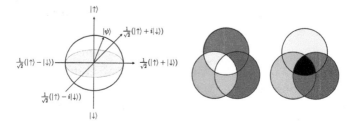

Fig. 2 Illustration of how qualia superpose to make new qualia. **Left**: The Bloch sphere illustrates the case of direction qualia. Summing and subtracting the quale "up" and the quale "down", we obtain the qualia "right" and "left", respectively. Similarly, if we make the same combinations with the imaginary unit i in front of "down" we obtain the qualia "front" and "back". Summing with generic complex amplitudes a and b (with $|a|^2 + |b|^2 = 1$), we get the quale corresponding to a generic direction, corresponding to a generic pure state $|\psi\rangle$. **Right**: Summations of two members of the three-colour basis *red*, *green*, and *blue* (RGB), and subtraction of two of the three colours *cyan*, *magenta* and *yellow* (CMY). As explained in the main text, these examples are only intended to illustrate the notion of a linear combination of quantum ontic states to make new ontic states corresponding to new qualia. A third case of superposition is that of sounds with precise frequency that combine with addition and subtraction into timbres and chords. Notice that all three cases fit the wave aspect of reality, not the particle one

with triple or quadruple or more entanglement in a factor of a tensor product, and more generally, every system is entangled, e.g.

$$\ldots \otimes (|a_1\rangle \otimes |b_1\rangle \otimes |c_1\rangle \otimes |d_1\rangle \pm |a_2\rangle \otimes |b_2\rangle \otimes |c_2\rangle \otimes |d_2\rangle) \otimes (|\uparrow\rangle \otimes |\bullet\rangle \pm |\downarrow\rangle \otimes |\circ\rangle) \otimes \ldots \tag{3}$$

We can see how in this fashion one can achieve new kinds of qualia whose number grows exponentially with the number of systems. In fact, the number of different ways of entangling N systems corresponds to the number of partitions of N into integers, multiplied by the number of permutations of the systems, and therefore it grows as

$N!\text{Exp}(\sqrt{2N/3})$, and this without considering the variable vectors that can be entangled!

Since qualia correspond to pure states of conscious systems, their Hilbert space coincides with the multidimensional Hilbert space of the system. Experimentally, one may be able to locate the systems in terms of neural patterns, but we will never be able to read the encoded information without destroying the person's experience, while at best gaining only a single complementary side of the qualia, out of exponentially many. In short:

- System identification is possible, but not the "experience" within.

The fact that it is possible to identify the information system proves that the identity of the observer/agent is public and is thus correlated with its "sense of self," which is instead private. This is a crucial requirement for a unified theory of consciousness and free will, namely that the observer/agent be identifiable both privately–from within and through qualia–and publicly–from without and through information. Within QT (or post-quantum OPTs), this is possible.

4 Evolution of Consciousness and Free Will

Since the quantum state of a conscious system must be pure at all times, the only way to guarantee that the evolving system state remains pure is for the evolution itself to be pure (technically it is *atomic*, i.e. namely its CP-map $\mathcal{T} = \sum_i T_i \cdot T_i^\dagger$ has a single Krauss operator T_{i_0}). Note that both states and effects are also transformations from trivial to non trivial systems and viceversa, respectively.

Table 1 Notations for the three kinds of ontic transformation (for the meaning of the symbols, see Table 2). For the list of axioms and the main theorems of quantum theory, see Tables 3 and 4

	Circuit symbol	Symbol	Map	Operator	Domain-codomain
transformation	A─[$\mathcal{T}$]─B	$\mathcal{T}$	$\mathcal{T} = T \cdot T^\dagger$	T	$\mathrm{Bnd}(\mathcal{H}_A \to \mathcal{H}_B)$
Effect	A─[α]	$(\alpha\|$	$\mathrm{Tr}[\cdot \|\alpha\rangle\langle\alpha\|]$	$\langle\alpha\|$	$\mathrm{Bnd}(\mathcal{H}_A \to \mathbb{C})$
State	(ψ)─A	$\|\psi)$	$\|\psi\rangle\langle\psi\|$	$\|\psi\rangle$	$\mathrm{Bnd}(\mathbb{C} \to \mathcal{H}_A)$

In Table 1 we report the theoretical representations of the three kinds of ontic transformations, including the special cases of state and effect.

Now consider the general scenario of a composite conscious system in an ontic state ω_t at time t, evolved by the one-step *ontic transformation* $\mathcal{O}^{(t,x_t)}_{F_t}$ with outcome F_t, depending on classical input x_t from the senses and memory[9]

$$(\omega_t)\!-\!\!{}^{A_t}\!\!\boxed{\mathcal{O}^{(t,x_t)}_{F_t}}\!\!-\!\!{}^{A_{t+1}} \;=:\; (\omega_{t+1})\!-\!\!{}^{A_{t+1}}. \qquad (4)$$

The *epistemic transformation* would be the sum of all ontic ones corresponding to all possible outcomes:

$$\mathcal{E}^{(t,x_t)} := \sum_{F_t} \mathcal{O}^{(t,x_t)}_{F_t}. \qquad (5)$$

The outcome F_t is a classical output, and we identify it with the *free will* of the experiencing system.

It is a probabilistic outcome that depends on the previous history of the qualia of the system. Its randomness is the quantum kind, which means that *it cannot be interpreted as lack of knowledge*, and, as such, *it is free*. Notice that both mathematically and literally the free will is the

---

[9] As we will see, memory is classical.

outcome of a transformation that corresponds to a change of experience of the observer/agent. The information conversion from quantum to classical can also take into account a stage of "knowledge of the will" corresponding to "intention/purpose", namely "understanding" of which action is taken.

We may need to provide a more refined representation of the one-step ontic transformation of the evolution in terms of a quantum circuit, for example:

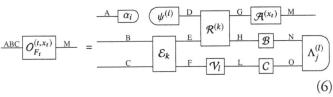

(6)

Generally for each time t we have a different circuit. In the example in circuit (6) we see that a single step can also contain states and effects, and the output system of the whole circuit (M in the present case) is generally different (not even isomorphic) to the input systems ABC. Following the convention used for the ontic transformation $\mathcal{O}_{F_t}^{(t,x_t)}$ the lower index is a random outcome,[10] and the upper index (t, x_t) is a parameter on which the transformation generally depends. Overall in circuit (6) we have free will $F_t = \{i, k, l, j\}$, whereas the transformations $\psi^{(i)}, \mathcal{R}^{(k)}, \mathcal{B}$, and $\mathcal{C}$ are deterministic (they have no lower index), and do not contribute to the free will; the same goes for the transformation $\mathcal{A}^{(x_t)}$ which depends on the sensorial input x_t. In circuit (6), we can see that *classical information is also at work*, since, e.g., the transformation $\mathcal{R}^{(k)}$ depends on the outcome k of the transformation $\mathcal{E}_k$. Similarly, the choice of test $\{\Lambda_j^{(l)}\}$ depends on the outcome l of $\mathcal{V}_l$, and the state

---

[10] The outcome is random for an observer other than the conscious system, for which it is instead precisely known.

$\psi^{(i)}$ depends on the effect outcome α_i. Each element of the circuit is ontic, i.e. atomic, and atomicity of sequential and parallel composition guarantee that the whole transformation is itself atomic. This means that its Krauss operator O_t can be written as the product of the Krauss operators of the component transformations as follows

$$O^{(t,x_t)}_{F_t} = \underbrace{\left(I_M \otimes \left\langle \Lambda_j^{(l)}\right|\right)}_{MNO \to M} \underbrace{\left(A^{(x_t)} \otimes B \otimes C\right)}_{GHL \to MNO} \underbrace{\left(R^{(k)} \otimes V_l\right)}_{DEF \to GHL} \underbrace{(|\psi_i\rangle\langle a_i| \otimes E_k)}_{ABC \to DEF},$$
(7)

where $\mathcal{O}^{(t,x_t)}_{F_t} = O^{(t,x_t)}_{F_t} \cdot O^{(t,x_t)\dagger}_{F_t}$. Note that in the expression for the operator $O^{(t,x_t)}_{F_t}$ in Eq. (7) the operators from the input to the output are written from the right to the left, the way we compose operators on Hilbert spaces. Moreover, the expression (7) of $O^{(t,x_t)}_{F_t}$ is not unique, since it depends on the choice of *foliation* of the circuit, namely the way you cover all wires with *leaves* to divide the circuit into input–output sections. For example, Eq. (7) would correspond to the foliation in Fig. 3.

A different foliation, for example, is the one reported in Fig. 4, corresponding to the expression for $O^{(t,x_t)}_{F_t}$

$$O^{(t,x_t)}_{F_t} = \underbrace{\left(I_C \otimes \left\langle \Lambda_j^{(l)}\right|\right)}_{MNO \to M} \underbrace{\left(A^{(x_t)} \otimes B \otimes I_O\right)}_{GHO \to MNO} \underbrace{\left(R^{(k)} \otimes CV_l\right)}_{DEF \to GHO} \underbrace{(|\psi_i\rangle \otimes E_k)}_{BC \to DEF} \underbrace{(\langle a_i| \otimes I_{BC})}_{ABC \to BC}$$
(8)

Recall that all operators (including *kets* and *bras* as operators from and to the trivial Hilbert space $\mathbb{C}$) are *contractions*, i.e., they have norm bounded by 1, corresponding to marginal probabilities not greater than 1. Thus, also $O^{(t,x_t)}_{F_t}$ is itself a contraction. Contractivity for an operator X can be conveniently expressed as

$$X \in \text{Bnd}(\mathcal{H}_{in} \to \mathcal{H}_{out}), \quad X^\dagger X \le P_{S_X} \le I_{in}, \quad P_{S_X} := \text{Proj Supp}\, X.$$
(9)

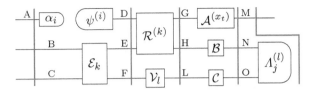

Fig. 3 Quantum circuit foliation corresponding to Eq. (7)

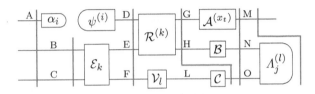

Fig. 4 Quantum circuit foliation corresponding to the expression for $O_{F_t}^{(t,x_t)}$ Eq. (8)

Now, since the evolution of consciousness must be atomic at all times, we can write a whole consciousness history as the product of the Krauss operators $O_t \equiv O_{F_t}^{(t,x_t)}$ at all previous times t

$$\Omega_t := O_t O_{t-1} \ldots O_1, \quad O_t := O_{F_t}^{(t,x_t)}, \qquad (10)$$

and apply the history operator to the wavevector of the initial ontic state-vector $|\omega_0\rangle$

$$|\omega_t\rangle = \Omega_t |\omega_0\rangle. \qquad (11)$$

The squared norm $||\omega_t||^2$ of the vector ω_t is the probability of the full *history of conscious states* $\{\omega_0, \omega_1, \omega_2, \ldots, \omega_t\}$ and also of the *free-will history* $\{F_1, F_2, \ldots, F_t\}$

$$p(\omega_0, \omega_1, \omega_2, \ldots, \omega_t) = p(\emptyset, F_1, \ldots, F_t) = ||\omega_t||^2 = ||\Omega_t \omega_0||^2. \qquad (12)$$

5 Memory

The ontic evolution of the consciousness state, though it maintains coherence, can keep very limited quantum memory of experience. The latter, due to contractivity of O, will go down very fast as $\|\Omega_t\|^2 \simeq \|O\|^{-2t}$, i.e., it will decrease exponentially with the number of time-steps. The quantum memory intrinsic in the ontic evolution instead works well as a *short-term buffer* to build up a fuller experience–e.g., of a landscape, or of a detailed object, or even to detect motion. How many qubits will make such a single-step buffer? The answer is: not so many.[11] Indeed, if we open our eyes for just a second to look at an unknown scene, and thereafter we are asked to answer binary questions, we would get only a dozen of right answers better than chance. Tor Nørretranders writes [31]: *The bandwidth of consciousness is far lower than the bandwidth of our sensory perceptors. ... Consciousness consists of discarded information far more than information present. There is hardly any information left in our consciousness.*[12]

---

[11] For example, the amount of visual information is significantly degraded as it passes from the eye to the visual cortex. Marcus E. Raichle says [29]: *Of the virtually unlimited information available from the environment only about 10^{10} bits/sec are deposited in the retina. Because of a limited number of axons in the optic nerves (approximately 1 million axons in each) only 6×10^6 bits/sec leave the retina and only 10^4 make it to layer IV of V1* [30, 31]. These data clearly leave the impression that visual cortex receives an impoverished representation of the world, a subject of more than passing interest to those interested in the processing of visual information [32]. Parenthetically, it should be noted that estimates of the bandwidth of conscious awareness itself (i.e. what we "see") are in the range of 100 bits/sec or less [30, 31].

[12] We believe that the inability to recall much information contained in one second of visual experience, when the actual experience is felt to be quite rich, should not be construed to diminish the importance of consciousness. In fact, experience is quantum while memory is classical, and although not much classical information appears to have been memorized, the actual experience has the cardinality of the continuum in Hilbert space. Consciousness is about

We can reasonably deduce that there is actually no room for long-term memory in consciousness, and we conclude that:

S3: Memory is classical. Only the short-term buffer to collect each experience is quantum.

By "short" we mean comparable to the time in which we collect the full experience, namely of the order of a second.

Transferring quantum experience to classical memory

If we have quantum experiences and classical memory, we need to convert information from quantum to classical in the memorization process, and from classical to quantum in the recollection process. The first process, namely transferring quantum experience to classical memory, must necessarily be incomplete, otherwise it would violate the quantum no-cloning theorem. The Holevo theorem [33] establishes that the maximal amount of classical information that can be extracted from a quantum system is a number of bits equal to the number of qubits that constitute the quantum system. Obviously, such maximal classical information is infinitesimal compared to the continuum of classical information needed to communicate a quantum state classically!

From a single measurement one can extract classical information about just one of the continuum of complementary aspects of the quantum state, e.g. along a given direction for

---

living the experience in its unfolding and understanding what is happening, so as to make the appropriate free-will decisions when necessary. Recalling specific information in full detail is unnecessary. Consciousness is focused on the crucial task of getting the relevant meaning contained in the flow of experience. Our scarce conscious memory of specific objects and relationships between objects should not be taken as an indication that consciousness is a "low bandwidth" phenomenon, but that what is relevant to consciousness may not be what the investigator believes should be relevant.

a spin. Moreover, the state after the measurement would necessarily be disturbed,[13] due to the *information-disturbance tradeoff*.[14] So, which measurement should one perform to collect the best classical information from a quantum system?

The answer is easy for a single qubit. Just perform the usual von Neumann measurement along a random direction! This has been proved in [35].[15] For higher dimensions it is in principle possible to generalise this result. However, the probability space turns out to be geometrically much

---

[13] For example, for the spin originally oriented horizontally and measured vertically *a la* von Neumann, the final state would be vertical up or down, depending on the measurement outcome.

[14] There are many ways of regarding the information-disturbance tradeoff, depending on the specific context and the resulting definitions of *information* and *disturbance*. The present case of *atomic measurements* with the "disturbance" defined in terms of the *probability of reversing the measurement transformation* has been analysed in the first part of [34]. In the same reference it is also shown that a reversal of the measurement would provide a contradicting piece of information which numerically cancels the information achieved from the original measurement, thus respecting the quantum principle of no information without disturbance.

[15] Consider the case of a single qubit realised with a particle spin. The usual observable of von Neumann corresponds to a measurement of the orientation of the spin along a given direction, e.g., *up* or *down* along the vertical direction, or *left* or *right* along the horizontal direction. But when we prepare a spin state (e.g., by Rabi techniques), we can put the spin in a very precise direction, e.g., pointing north-east along the diagonal from south-west to north-east, and, indeed if we measure the spin along a parallel direction we find the spin always pointing north-east! So the spin is indeed (ontically!) pointing north-east along the same diagonal! Now it would be legitimate to ask: *how about measuring the direction of the spin itself?* This can be done–not exactly, but optimally–using a continuous observation tests constituted by a continuum of effects (what is usually Positive Operator Valued Measure or POVM). However, it turns out that the measured direction is a fake, since such a quantum measurement with a continuous set of outcomes is realised as a continuous random choice of a von Neumann measurement [35]. In conclusion the optimal measurement of the spin direction is realised by a standard Stern–Gerlach experiment in which the magnetic field is randomly oriented! The same method can be used to achieve an informationally complete measurement to perform a quantum tomography [36] of the state by suitably processing the outcome depending on the orientation of the spin measurement.

more complicated than a sphere,[16] and a scheme for random choice is unknown. For this reason we can consider the larger class of observation tests referred to as *informationally complete* ("infocomplete"),[17] which, unlike the case of the optimal measurement of spin direction, can be taken as discrete and with a finite number of outcomes. Such a measurement would correspond to the following epistemic transformation:

$$\mathcal{M} \in \text{Trn}_1(A \to B), \quad \mathcal{M} = \sum_{j \in J} M_j \rho M_j^\dagger, \quad |J| \geq \dim(\mathcal{H}_A)^2, \tag{13}$$

which is the coarse graining of the *infocomplete* quantum test $\{\mathcal{M}_j\}_{j \in J}$, with $\mathcal{M}_j = M_j \cdot M_j^\dagger$, where $\{|M|_j^2\}_{j \in J}$ is an *infocomplete observation test*. "Infocomplete" means that, in the limit of infinitely many usages over the same reprepared state, the measurement allows one to perform a full tomography [36] of the state.

The infocomplete measurement (or the random observables seen before) is better suited to extract classical information from the quantum buffer for long-term memory, since it does not privilege any particular observable. It is also likely that the conscious act of memorising an experience may be achieved by actually *repreparing the ontic state multiple times in the quantum buffer*, and performing the infocomplete test multiple times, thus with the possibility of memorising a (possibly partial) tomography of the state. Notice that, generally, the ontic state $M_j \rho M_j^\dagger$ for a given

---

[16] The convex set of deterministic states for a *qutrit* (i.e. $d = 3$), defined by algebraic inequalities, has eight dimensions, and has a boundary made of a continuum of balls.

[17] What in OPT language is called observation test is the same as what in the quantum information literature is called a discrete POVM. An observation test is infocomplete for system A when it spans the linear space of effects $\text{Eff}_\mathbb{R}(A)$.

j still depends on ρ. However, the test will disturb it, and the less disturbance, the smaller the amount of classical information that can be extracted [34].[18]

Transferring classical memory to quantum experience

We have seen that an infocomplete measurement is needed to (approximately) store an experience in the classical long-term memory. By definition, the recovery of an experience requires a reproduction of it, meaning that the corresponding ontic state is (approximately) reprepared from the classical stored data. The memory, being classical, will be read without disturbance, thus left available for following recollections. In order to transfer classical to quantum information, methods of state-preparation have been proposed in terms of quantum circuit schemes [38].

A possible benchmark for the memory store-and-recall process is the maximal fidelity achievable in principle with a *measure-and-reprepare* scheme that optimises the fidelity between the experienced state and the recalled one. We will consider $M \geq 1$ input copies of the same state to quantify *attention* to the experience. The optimal *fidelity*[19] between the experienced state and the recalled one

---

[18] Particularly symmetric types of measurements are those made with a SIC POVM (symmetric informationally complete POVM) [37], where $M_j = |\psi_j\rangle\langle\psi_j|$ are d^2 projectors onto pure states with equal pairwise fidelity.

$$|\langle\psi_j|\psi_k\rangle|^2 = \frac{d\delta_{jk} + 1}{d + 1}. \qquad (14)$$

The projectors $|\psi_j\rangle\langle\psi_j|$ defining the SIC POVM in dimension d form a $(d^2 - 1)$-dimensional regular simplex in the space of Hermitian operators.

[19] The fidelity F between two pure states corresponding to state vectors $|\psi\rangle$ and $|\varphi\rangle$, respectively, is defined as $|\langle\varphi|\psi\rangle|^2$..

averaged over all possible experiences (i.e., input states) is given by [39]

$$F(A, M, d) = \frac{M+1}{M+d}, \quad (15)$$

where d is the dimension of the experiencing system[20].

Such an upper bound for fidelity may be used to infer an effective dimension d of the system involved in consciousness, e.g. in highly focused restricted experiences like those involved in masking conscious perception [40].

Information transfer from body to consciousness and vice versa

We would expect most of the operation of the human body to be automatic and use classical information that is never translated into quantum information to be experienced by consciousness. The portion of the classical information produced by the body that needs to be converted into quantum should only be the salient information that

---

[20] The optimal fidelity in Eq. (15) is achieved by an observation test with atomic effects $|\Phi_i\rangle\langle\Phi_i|$ with

$$|\Phi_i\rangle = \sqrt{w_i d_M}|\psi_i\rangle^{\otimes M}, \quad d_M = \binom{M+d-1}{d-1} \quad (16)$$

where $\{|\psi_i\rangle\}_{i=1}^{A}$ are pre-specified pure state vectors, and $\{w_i\}_{i=1}^{A}$ is a probability vector satisfying the identity.

$$\sum_{i=1}^{A} w_i |\psi_i\rangle\langle\psi_i|^{\otimes M} = \left\langle |\psi\rangle\langle\psi|^{\otimes M} \right\rangle. \quad (17)$$

Here the notation $\langle \ldots \rangle$ denotes averaging over the prior of pure-state vectors, taking as the prior the Haar measure on the unit sphere in $\mathbb{C}^d$.

supports the qualia perception and comprehension necessary to "live life" and make appropriate free-will choices. The amount of quantum information to be translated into classical for the purpose of free-will control of the body top-down should be relatively small.

6 On the Combination Problem

The *combination problem* concerns the issue about *whether and how the fundamental conscious minds come to compose, constitute, or give rise to some other, additional conscious mind* [41]. By definition, the problem becomes crucial for panpsychism: if consciousness is everywhere, what is the criterion for selecting novel conscious individuals? Is the union of two conscious beings a conscious being? If this is true, then any subset of a conscious being can also be a conscious being. The present theoretical approach provides a precise individuation criterion. The criterion derives from principle P3 about the purity of quantum conscious states and, consequently, the need for ontic transformations. Let's see how it works.

It is reasonable to say that *an individual is defined by the continuity of its experience.* Such a statement may be immediately obvious to some readers. However, for those who may not agree, we propose a thought experiment.

Consider a futuristic "quantum teleportation experiment", meaning that the quantum state of a system is replaced by that of a remote isomorphic system.[21] *The matter (electrons, protons, etc.) of which a teleported person is composed is available at the receiver location and is in-principle*

---

[21] Teleportation will need the availability of shared entanglement and classical communication, and technically would use a Bell measurement at the sender and a conditioned unitary transformation at the receiver [42].

Fig. 5 The combination problem. Stated generally, the problem is about how fundamental conscious minds come to compose, constitute, or give rise to some further conscious mind. The ontic-state principle P1 provides a partial criterion to exclude some situations, e.g. (figure on the left), two entangled systems cannot separately be conscious entities, since each one is in a marginal state of an entangled one, hence it is necessarily mixed. On the other hand, (figure on the right), if two systems are in a factorized pure state, each system is in an ontic state, and they are two single individuals. They will then remain so depending on their following interactions (see Fig. 6)

indistinguishable from the same matter at the transmission location: indeed, it is only its quantum state that is teleported. The resulting individual would be the same as the original one, including his thoughts and memory.[22]

The above thought experiment suggests the following *individuation criterion* within our theoretical approach:

S4: A conscious mind is a composite system in an ontic state undergoing an ontic transformation, with no subsystem as such.

In Figs. 5 and 6 we illustrate the use of the criterion in two paradigmatic cases. Here we emphasise that the role of interactions is crucial for the criterion.

---

[22] Of course, this would violate the no-cloning theorem, if at the transmission point the original individual were not destroyed. Indeed, according to the quantum information-disturbance tradeoff [34], teleportation cannot even make a bad copy, leaving the original untouched.

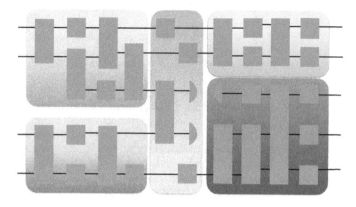

Fig. 6 **The combination problem.** The general individuation criterion in statement S4 requires full quantum coherence, namely the ontic nature of both states and transformations. In the case depicted in this figure, every transformation (including effects and states as special cases) is pure, and we suppose each multipartite transformation is not factorizable. Then the first two boxes on the left represent two separate individuals. The box in the middle merges the two individuals into a single one, whereas the immediately subsequent effects convert quantum to classical information, and separate the single individual into the original separate ones once again. Notice that a merging of two individuals necessarily involves a quantum interaction

This section has described what is necessary to form a combination of conscious entities, thus removing the difficulties encountered with panpsychist models based on classical physics. The key idea is that the states of the combining systems and their transformations be ontic and that such systems interact quantumly. These requirements assure that the combined entity is also in a pure state and that *none of its subsets are conscious.* Clearly, the interactions between the conscious entity and the environment (including other entities) can only be classical, otherwise a larger entity would be created and would persist for as long as a disentangling transformation does not occur, as in the case illustrated in Fig. 6.

7 Experimentability and Simulability

Proposing feasible experiments regarding the quantum nature of consciousness is a very exciting challenge. In principle, quantumness could be established through experimental demonstration of one of the two main non-classical features of quantum theory: *nonlocality* and *complementarity*. The two notions are not independent, since in order to prove nonlocality, we need complementary observations, in addition to shared entanglement.

In order to prove nonlocality of consciousness, we need measurements at two separate points sufficiently far apart to exclude causal connection, and such a requirement is very challenging, since it demands very fast measurements, considering that, for a distance of 3 cm between the measurements points, a time-difference of a tenth of a nanosecond is sufficient to have signalling.

Regarding nonlocality together with complementarity, we speculate that they may together be involved in 3D vision, and take the opportunity to suggest that such a line of research may in any case be a fruitful field of experimental research on consciousness.[23] For example, a genuine case of complementarity in 3D experience occurs as a result of looking at *Magic Eye* images published in a series of books [44]. These images feature *autostereograms*, which allow most people to see 3D images by focusing on 2D patterns that seem to have nothing in common

---

[23] A paradigmatic case of superposition between incompatible 3D views is that of the Necker cube, which some authors regard as a neuro-physiological transformation leading to perceptual reversal controlled by the principles of quantum mechanics [43]. However, the 3D experience of the Necker cube does not require binocular disparity since monocular vision also produces the same effect.

with the 3D image that emerges from them. The viewer must diverge or converge his eyes in order to see a hidden three-dimensional image within the patterns. The 3D image that shows up in the experience is like a glassy object that, depending on convergence or divergence of the eyes, shows up as either concave or convex. Clearly the two alternative 3D views–convex and concave–are truly complementary experiences, and each experience has an intrinsic wholeness.

Regarding complementarity alone, speculative connections with contrasting or opposite dimensions of human experience have been considered in the literature, e.g. "analysis" versus "synthesis", and "logic" versus "intuition" [45]. Here, however, we are interested in *complementary in experiences possibly reproducible in different individuals*, of the kind of *gedanken-experiments a la Heisenberg*, e.g. one experience incompatible with another and/or disturbing each other. Having something more than two complementary observables, namely an informationally complete set of measurements such as all three Pauli matrices for a single consciousness qubit, would allow us to carry out a quantum tomography [36] of the qubit state, assuming that this is reprepared many times, e.g. through intense prolonged attention. To our knowledge, the feasibility assessment of this kind of experiment is of similar difficulty to a nonlocality experiment.

Finally, apart from proving the quantumness of experience, we can at least experimentally infer some theoretical parameters in the quantum theoretical approach consistent with observations. This is the case e.g. of the experiment already mentioned when discussing the upper bound (15) in memory-recall fidelity, which can be used to inferring an *effective dimension d* of the system involved in consciousness in highly focused restricted experiences, such as masking conscious perception [40]. We believe that memory-recall experiments versus variables such as attention, memory-recovery delay, and variable types of qualia, may be helpful to make a preliminary mapping of the dimensionality of the spaces of the conscious systems involved.

We conclude with a few considerations about the feasibility of a simulation of a conscious process like the ones proposed here. As already mentioned, due to the purity of the ontic process, a simulation just needs multiplications of generally rectangular matrices, which for matrices with the same dimension d is essentially a $\Theta(d^3)$ process. For sparse matrices (as is often the case in a quantum simulation) the process can be speeded up considerably. However, to determine the probability distribution of each ontic step, one needs to evaluate the conditional probability for the output state of the previous step, and this involves the multiplication for all possible outcomes (the free will) of the corresponding Kraus operator of the last ontic transformation. With a RAM of the order of Gbytes one could definitely operate with a dozen qubits at a time. This should be compared with 53 qubits of the Google or IBM quantum computer, the largest currently available, likely to share classical information in tandem with a large classical computer. However, it is not excluded that some special phenomena could already be discovered/analyzed with a laptop.

8 Philosophical Implications

We have presented a theory of consciousness, based on principles, assumptions, and key concepts that we consider crucial for the robustness of the theory and the removal of the limitations of most panpsychist theories [41]. We believe that conferring inner reality and agency on quantum systems in pure quantum states, with conversion of information from classical to quantum and vice versa, is unprecedented, with major philosophical and scientific consequences. In the present approach free will and consciousness go hand in hand, allowing a system to act on the basis of its qualia experience by converting quantum to classical information, and thus giving causal power to subjectivity–something that until now has been considered highly controversial, if not impossible.

The theory provides that a conscious agent may intentionally convert quantum information into a specific piece of classical information to express its free will, a classical output that is in principle unpredictable due to its quantum origin. The theory would be incoherent without the identification of the conscious system in terms of purity and inseparability of the quantum state, which is identified with the systems experience. The purity of non-deterministic quantum evolution identifies consciousness with agency through its outcome.

Metaphysically, the proposed interpretation that a pure, non-separable quantum state is a state of consciousness could be turned on its head by assuming the ontology of consciousness and agency as primary, whereupon physics would be emergent from consciousness and agency. This same interpretation would then consider classical physics to be the full reification (objectification) of quantum reality as quantum-to-classical agency corresponding to the free will of conscious entities existing entirely in the quantum realm. The ontology derived by accepting consciousness as fundamental would be that objectivity and classical physics supervene on quantum physics, quantum physics supervenes on quantum information, and quantum information supervenes on consciousness. If we were to accept this speculative view, physics could then be understood as describing an open-ended future not yet existing because the free-will choices of the conscious agents have yet to be made. In this perspective, we, as conscious beings, are the co-creators of our physical world.

We do so individually and collectively, instant after instant and without realizing it, by our free-will choices.

Appendices about general OPTs

In these appendices we provide a general operational probabilistic framework for future post-quantum explorations of a theory of consciousness. We give here a short illustration of what constitutes an operational probabilistic theory (OPT), with quantum theory and classical theory being the most relevant instances. As the reader may appreciate, OPT provides a framework that is much richer, more general, flexible, and more mathematically rigorous that other theoretical frameworks, such as the causal approach of Tononi [46].

The operational probabilistic theory (OPT) framework

It is not an overstatement to say that the OPT framework represents a new Galilean revolution for the scientific method. In fact, it is the first time that a theory-independent set of rules is established on how to build a theory in physics and possibly in other sciences. Such rules constitute what is called the *operational framework*. Its rigour is established by the simple fact that the OPT is just "metamathematics", since it is a chapter of *category theory* [47, 48]. To be precise the largest class of OPTs corresponds to a *monoidal braided category*. The fact that the same categoric framework is used in computer science [49–53] gives an idea of the thoroughness and range of applicability of the rules of the OPT.

In several seminal papers, Lucien Hardy [54–56] introduced a heuristic framework that can be regarded as a forerunner of the OPT, which made its first appearance in [8,

9] and was soon connected to the categorical approach in computer science [20–25]. As already mentioned, both QT and CT are OPTs [10], but one can build up other OPTs, such as variations of QT, e.g., fermionic QT [57], or QT on real Hilbert space [10], or QT with only qubits, but also CT with entanglement [58] or without local discriminability [59]. Other toy-theories, such as the PR-Boxes, are believed to be completable to OPTs (see e.g., [60]).

The connection of OPT with computer science reflects the spirit of the OPT, which was essentially born on top of the new field of quantum information. Indeed, the OPT framework is the formalization of the rules for building up quantum circuits and for associating a joint probability to them: in this way, the OPT literally becomes an extension of probability theory.

The general idea

How does an OPT work? It associates to each joint probability of multiple events a *closed directed acyclic graph* (CDAG) of input–output relations, as in the Fig. 7. Each *event* (e.g., $\mathcal{E}_m$ in the figure) is an element of a complete *test* ($\{\mathcal{E}_m\}_{m \in M}$ in the figure) with normalized marginal probability $\sum_{m \in M} p(\mathcal{E}_m) = 1$. The graph tells us that the marginal probability distribution of any set of tests still depends on the marginalized set, e.g., the marginal probability distribution of test $\{\mathcal{E}_m\}_{m \in M}$ depends on the full graph of tests to which it is connected, although it has been partially or fully marginalised. As a rule, disconnected graphs (like γ_1 and γ_2 in Fig. 7) are statistically independent, i.e., their probability distributions factorize.

The oriented wires denoting output-input connections between the tests (labeled by Roman letters in Fig. 7), are the so called *systems* of the theory.

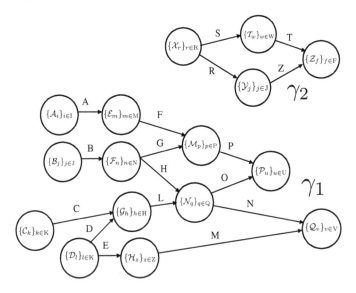

Fig. 7 An OPT associates a closed directed acyclic graph (CDAG) of output-input relations with each joint probability distribution of multiple tests/events (see text). As a rule, unconnected graph components are statistically independent, e.g., in the case in the figure, one has $p(\mathcal{A}_i, \mathcal{E}_m, \ldots, \mathcal{X}_r, \ldots | \gamma_1 \cup \gamma_2) = p(\mathcal{A}_i, \mathcal{E}_m, \ldots | \gamma_1,) p(\mathcal{X}_r, \ldots | \gamma_2)$

A paradigmatic quantum example

A paradigmatic example of an OPT graph is given in Fig. 8. There, we have a source of particles all with spin up (namely in the state $\rho = |\uparrow\rangle\langle\uparrow|$). At the output we have two von Neumann measurements[24] in cascade, the first one Σ_α measuring σ_α, and the second one Θ_β measuring σ_β, where α and β can assume either of the two values x, z. The setup is represented by the graph shown in the figure, where A represents the system corresponding to

---

[24] A von Neumann measurement of e.g. σ_z has two outcomes "up" and "down", and the output particle will be in the corresponding eigenstate of σ_z.

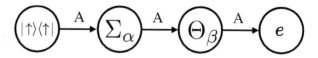

Fig. 8 **A simple OPT graph.** A paradigmatic example for the sake of illustration and for motivation (see text)

the particle spin, e the deterministic test that simply discards the particle, and the two tests $\Sigma_\alpha, \Theta_\beta$ ($\alpha, \beta = x, z$) the two von Neuman measurements. Now, clearly, for $\alpha = z$ one has the marginal probability distribution $p(\Sigma_z) = (1,0)$ and $p(\Sigma_x) = \left(\frac{1}{2}, \frac{1}{2}\right)$ independently of the choice of the test Θ_β. On the other hand, for the second test one has marginal probability $p(\Theta_x) = (1,0)$ for Σ_z and $p(\Theta_x) = \left(\frac{1}{2}, \frac{1}{2}\right)$ for Σ_x. We conclude that *the marginal probability of Σ_α is independent of the choice of the test Θ_β, whereas the marginal probability of Θ_β depends on the choice of Σ_α.* Thus, the marginal probability of Θ_β generally depends on the choice of Σ_α, and this concept goes beyond the content of joint probability, and needs the OPT graph. Theoretically, we conclude that there is "something flying" from test Σ_α to test Θ_β (although we cannot see it!): this is what we describe theoretically as a spinning particle! This well illustrates the notion of *system*: a theoretical connection between tested events.

A black-box approach

Finally, the OPT is a *black-box approach*, where each test is described by a mathematical object which can be "actually achieved" by a very specific physical device. However, nothing prevents us from providinge a more detailed OPT realisation of the test, e.g. as in Fig. 9.

Notice that although any graph can be represented in 2D (e.g., using the crossing of wires), one can more suitably design it in 3D (2D + in–out), as in Fig. 10.

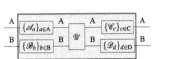 

Fig. 9 OPT finer and coarser descriptions. The OPT is a *black-box approach*. It can be made more or less detailed, as in in the box on the left, and even at so fine a level that it is equivalent to a field-theoretical description, as on the right

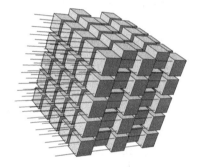

Fig. 10 An (open) DAG whose topology is suitably representable in 3D

The OPT and the goal of science

One can soon realise that the OPT framework allows one to precisely express the most general goal of science, namely to connect objective facts happening (events), devising a theory of such "connections" (systems), and thus allowing one to make predictions for future occurrences in terms of joint probabilities of events depending on their connections.

One of the main methodologically relevant features of the OPT is that it perfectly distinguishes the "objective datum" from the "theoretical element". What is *objective* are the tests that are performed and the outcome of each test. What is *theoretical* is the graph of connections between the tests, along with the mathematical representations of both systems and tests. The OPT framework dictates the rules that the mathematical description should satisfy, and the specific OPT gives the particular mathematical representation of systems and tests/events and of their compositions (in sequence and in parallel) to build up the CDAG.

The OPT and the scientific method

One of the main rules of the scientific method is to have a clearcut distinction between what is experimental and what is theoretical. Though this would seem a trivial statement, such a confusion often happens to be a source of disagreement between scientists. Though the description of the apparatus is generally intermingled with theoretical notions, the pure experimental datum must have a conventionally defined "objectivity status", corresponding to "openly known" information, namely shareable by any number of different observers. Then both the theoretical language and the framework must reflect the theory–experiment distinction, by indicating explicitly which notions are assigned the objectivity status. Logic, with the Boolean calculus of events, is an essential part of the language, and probability theory can be regarded as an extension of logic, assigning probabilities to events. The notion that is promoted to the objectivity status is that of "outcome of a test", announcing which event of a given test has occurred. The OPT framework thus represents an extension of probability theory, providing a theoretical connectivity between events, the other theoretical

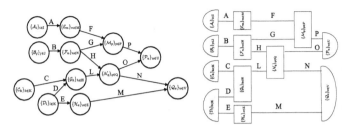

Fig. 11 Equivalence between the CDAG and the quantum information circuit or, equivalently, any run diagram of a program

ingredients being the mathematical descriptions of systems and tests.

The OPT as a general information theory

We see immediately that a CDAG is exactly the same graph of a quantum circuit as the one drawn in quantum information science. The quantum circuit, in turn, can be interpreted as the run-diagram of a program, where each test represents a subroutine, and the wires represent the registers through which the subroutines communicate data. Indeed, the OPT can be regarded as the proper framework for information science in general (Fig. 11).

For a recent complete presentation of the OPT framework, the reader is referred to the work [9, 10] or the more recent and thorough presentation [19].

Notation and abbreviations

(Table 2)

Table 2 Notation, special-cases corollaries, and common abbreviations

$\text{Bnd}^+(\mathcal{H})$	bounded positive operators over $\mathcal{H}$
$\text{CP}_\leqslant$	trace-non increasing completely positive map
$\text{CP}_=$	trace-preserving completely positive map
$\mathcal{H}$	Hilbert space over $\mathbb{C}$
$\text{Cone}(S)$	conic hull of S
$\text{Cone}_{\leqslant 1}(S)$	convex hull of $\{S \cup 0\}$
$\text{Conv}(S)$	convex hull of S
$\text{Eff}(A)$	set of effects of system A
$\text{Eff}_1(A)$	set of deterministic effects of system A
$\text{Mrkv}_\leqslant$	normalization-non-increasing right-stochastic Markov matrices
Mrkv_1	normalization-preserving right-stochastic Markov matrices
$\text{Prm}(n)$	$n \times n$ permutation matrices
$(\mathbb{R}^n)^+_{\leqslant 1}$	$\{\mathbf{x} \in \mathbb{R}^{n_A} \,|\, \mathbf{x} \geq 0, \mathbf{x} \leqslant 1\}$: simplex $\mathbf{S}^{n_A+1}$
$(\mathbb{R}^n)^+_1$	$\{\mathbf{x} \subset \mathbb{R}^{n_A} \,|\, \mathbf{x} \geq 0, \|\mathbf{x}\|_1 = 1\}$: simplex $\mathbf{S}^{n_A}$
$\text{St}(A)$	set of states of system A
$\text{St}_1(A)$	set of deterministic states of system A
$T(\mathcal{H})$	trace-class operators over $\mathcal{H}$
$T^+(\mathcal{H})$	trace-class positive operators over $\mathcal{H}$
$T^+_{\leqslant 1}(\mathcal{H})$	positive sub-unit-trace operators over $\mathcal{H}$
$T^+_{=1}(\mathcal{H})$	positive unit-trace operators over $\mathcal{H}$
$\text{Trn}(A \to B)$	set of transformations from system A to system B
$\text{Trn}_1(A \to B)$	set of deterministic transformations from system A to system B
$\mathbb{U}(\mathcal{H})$	unitary group over $\mathcal{H}$
Special-case corollaries	
	$T(\mathbb{C}) = \mathbb{C}, \; T^+(\mathbb{C}) = \mathbb{R}^+, \; T^+_{\leqslant 1}(\mathbb{C}) = [0,1], \; T^+_{=1}(\mathbb{C}) = \{1\}$
	$\text{CP}(T(\mathcal{H}) \to T(\mathbb{C})) = P(T(\mathcal{H}) \to T(\mathbb{C})) = \{\text{Tr}[\cdot E], E \in \text{Bnd}^+(\mathcal{H})\}$
	$\text{CP}(T(\mathbb{C}) \to T(\mathcal{H})) = P(T(\mathbb{C}) \to T(\mathcal{H})) = T^+(\mathcal{H})$
	$\text{CP}_\leqslant(T(\mathbb{C}) \to T(\mathcal{H})) \equiv T^+_{\leqslant 1}(\mathcal{H})$
	$\text{CP}_\leqslant(T(\mathcal{H}) \to T(\mathbb{C})) \equiv \{\epsilon(\cdot) = \text{Tr}[\cdot E], 0 \leqslant E \leqslant I\}$
	$\text{Mrkv}_\leqslant(n, 1) = (\mathbb{R}^n)^+_{\leqslant 1}$
	$\text{Mrkv}_1(n, 1) = (\mathbb{R}^n)^+_{=1}$
Abbreviations	
CT	Classical theory
OPT	Operational probabilistic theory
PR boxes	Popescu-Rohrlich boxes
QT	Quantum theory

Quantum theory

A minimal mathematical axiomatisation of quantum theory as an OPT is given in Table 3. For an OPT we need to provide the mathematical description of systems, their

Table 3 **Mathematical axiomatisation of quantum theory.** As given in the table, in quantum theory we associate a Hilbert space over the complex field $\mathcal{H}_A$ with each system A. We associate the tensor product of Hilbert spaces $\mathcal{H}_{AB} = \mathcal{H}_A \otimes \mathcal{H}_B$ with the composition of systems A and B. Transformations from A to B are described by trace-nonincreasing completely positive (CP) maps from traceclass operators on $\mathcal{H}_A$ to traceclass operators on $\mathcal{H}_B$. Special cases of transformations are those with trivial input system I corresponding to states whose trace is the preparation probability, the latter providing an efficient Born rule from which one can derive all joint probabilities of any combination of transformations. Everything else is simply special-case corollaries and one realisation theorem: these are reported in Table 4

Quantum theory		
system	A	$\mathcal{H}_A$
system composition	AB	$\mathcal{H}_{AB} = \mathcal{H}_A \otimes \mathcal{H}_B$
transformation	$\mathcal{T} \in \mathrm{Trn}(A \to B)$	$\mathcal{T} \in \mathrm{CP}_{\leqslant}(\mathrm{T}(\mathcal{H}_A) \to \mathrm{T}(\mathcal{H}_B))$
Born rule	$p(\mathcal{T}) = \mathrm{Tr}\,\mathcal{T}$	$\mathcal{T} \in \mathrm{Trn}(I \to A)$

composition, and transformations from one system to another. Then all rules for compositions of transformations and their respective systems are provided by the OPT framework. The reader who is not familiar with such a framework can simply use the intuitive construction of quantum circuits. In Table 4 we report the main theorems following from the axioms. The reader interested in the motivations for the present axiomatization is referred to [61].

Table 4 Corollaries and a theorem of quantum theory, starting from the axiomatization of Table 3. The first corollary states that, in order to satisfy the composition rule IA = AI = A, the trivial system I must be associated with the one-dimensional Hilbert space $\mathcal{H}_I = \mathbb{C}$, since it is the only Hilbert space which trivializes the Hilbert space tensor product. The second corollary states that the reversible transformations are the unitary ones. The third corollary states that the deterministic transformations are the trace-preserving ones. Then the fourth and fifth corollaries give the composition of transformations in terms of compositions of maps. We then have four corollaries about states: (1) states are transformations starting from the trivial system and, as such, are positive operators on the system Hilbert space, having trace bounded by one; (2) the deterministic states correspond to unit-trace positive operator; (3) the states of the trivial system are just probabilities; (4) the only trivial system deterministic state is the number 1. We then have two corollaries for effects, as special cases of transformation toward the trivial system: (1) the effect is represented by the partial trace over the system Hilbert space of the multiplication with a positive operator bounded by the identity over the system Hilbert space; (2) the only deterministic effect is the partial trace over the system Hilbert space. Finally, we have the realization theorem for transformations in terms of unitary interaction $\mathcal{U} = U \cdot U^\dagger$ with an environment F and a projective effect test $\{\mathcal{P}_i\}$ over environment E, with $\mathcal{P}_i = P_i \cdot P_i$, $\{P_i\}$ being a complete set of orthogonal projectors

Quantum theorems			
Trivial system		I	$\mathcal{H}_I = \mathbb{C}$
Reversible transf.		$\mathcal{U} = U \cdot U^\dagger$	$U \in U(\mathcal{H}_A)$
Determ. transformation		$\mathcal{T} \in \text{Trn}_1(A \to B)$	$\mathcal{T} \in \text{CP}_{\leqslant 1}(T(\mathcal{H}_A) \to T(\mathcal{H}_B))$
Parallel composition	$\mathcal{T}_1 \in \text{Trn}(A \to B)$, $\mathcal{T}_2 \in \text{Trn}(C \to D)$		$\mathcal{T}_1 \otimes \mathcal{T}_2$
Sequential composition	$\mathcal{T}_1 \in \text{Trn}(A \to B)$, $\mathcal{T}_2 \in \text{Trn}(B \to C)$		$\mathcal{T}_2 \mathcal{T}_1$
States		$\rho \in \text{St}(A) \equiv \text{Trn}_1(I \to A)$	$\rho \in T^+_{\leqslant 1}(\mathcal{H}_A)$
		$\rho \in \text{St}_1(A) \equiv \text{Trn}_1(I \to A)$	$\rho \in T^+_{=1}(\mathcal{H}_A)$
		$\rho \in \text{St}(I) \equiv \text{Trn}(I \to I)$	$\rho \in [0,1]$
		$\rho \in \text{St}_1(I) \equiv \text{Trn}_1(I \to I)$	$\rho = 1$
Effects		$\epsilon \in \text{Eff}(A) \equiv \text{Trn}(A \to I)$	$\epsilon(\cdot) = \text{Tr}_A[\cdot E]$, $0 \leqslant E \leqslant I_A$
		$\epsilon \in \text{Eff}_1(A) \equiv \text{Trn}_1(A \to I)$	$\epsilon = \text{Tr}_A$
Transformations as unitary interaction + von Neumann Lüders		A—[$\mathcal{T}_i$]—B = A—⎡$\mathcal{U}$⎤—B (σ)—F ⎣ ⎦E—$\mathcal{P}_i$	$\mathcal{T}_i \rho = \text{Tr}_E[U(\rho \otimes \sigma)U^\dagger(I_B \otimes P_i)]$

Classical theory

Table 5 Mathematical axiomatisation of classical theory.. With each system A, we associate a real Euclidean space $\mathbb{R}^{n_A}$. With composition of systems A and B, we associate the tensor product spaces $\mathbb{R}^{n_A} \otimes \mathbb{R}^{n_B}$. Transformations from system A to system B are described by substochastic Markov matrices from the input space to the output space. The rest are simple special-case corollaries: these are reported in Table 6

Classical theory		
System	A	$\mathbb{R}^{n_A}$
System composition	AB	$\mathbb{R}^{n_{AB}} = \mathbb{R}^{n_A} \otimes \mathbb{R}^{n_B}$
Transformation	$\mathcal{T} \in \mathrm{Trn}(A \to B)$	$\mathcal{T} \in \mathrm{Mrkv}_{\leq}(\mathbb{R}^{n_A}, \mathbb{R}^{n_B})$

Table 6 Main theorems of classical theory, starting from axioms in Table 5. The first corollary states that, in order to satisfy the composition rule $IA = AI = A$, the trivial system I must be associated with the one-dimensional Hilbert space $\mathbb{R}$, since it is the only real linear space that trivialises the tensor product. The second corollary states that the reversible transformations are the permutation matrices. The third states that transformations are substochastic Markov matrices. The fourth states that the deterministic transformations are stochastic Markov matrices. Then the fifth and sixth corollaries give the composition of transformations in terms of composition of matrices. We then have four corollaries about states: 1) states are transformations starting from the trivial system and as such are sub-normalized probability vectors (vectors in the positive octant with sum of elements bounded by one; 2) the deterministic states correspond to normalised probability vectors; 3) the case of the trivial output system correspond to just probabilities; 4) the only trivial output system deterministic state is the number 1. We then have two corollaries for effects, as special cases of transformation to the trivial system: 1) the effect is represented by the scalar product with a vector with components in the unit interval; 2) the only deterministic effect is the scalar product with the vector with all unit components

Classical theorems		
Trivial system	I	$\mathcal{H}_I = \mathbb{R}$
Reversible transformations	$\mathcal{P}$	$\mathcal{P} \in \mathrm{Prm}(n_A)$
Transformation	$\mathcal{T} \in \mathrm{Trn}_{\leq}(A \to B)$	$\mathcal{T} \in \mathrm{Mrkv}_{\leq}(\mathbb{R}^{n_A}, \mathbb{R}^{n_B})$
Determ. transformation	$\mathcal{T} \in \mathrm{Trn}_1(A \to B)$	$\mathcal{T} \in \mathrm{Mrkv}_1(\mathbb{R}^{n_A}, \mathbb{R}^{n_B})$
Parallel composition	$\mathcal{T}_1 \in \mathrm{Trn}(A \to B), \mathcal{T}_2 \in \mathrm{Trn}(C \to D)$	$\mathcal{T}_1 \otimes \mathcal{T}_2$
Sequential composition	$\mathcal{T}_1 \in \mathrm{Trn}(A \to B), \mathcal{T}_2 \in \mathrm{Trn}(B \to C)$	$\mathcal{T}_2 \mathcal{T}_1$
states	$\mathbf{x} \in \mathrm{St}(A) \equiv \mathrm{Trn}(I \to A)$	$\mathbf{x} \in (\mathbb{R}^{n_A})^+_{\leq 1}$
	$\mathbf{x} \in \mathrm{St}_1(A) \equiv \mathrm{Trn}_1(I \to A)$	$\mathbf{x} \in (\mathbb{R}^{n_A})^+_{=1}$
	$p \in \mathrm{St}(I) \equiv \mathrm{Trn}(I \to I)$	$p \in [0, 1]$
	$p \in \mathrm{St}_1(I) \equiv \mathrm{Trn}_1(I \to I)$	$p = 1$
effects	$\epsilon \in \mathrm{Eff}(A) \equiv \mathrm{Trn}(A \to I)$	$\epsilon(\cdot) = \cdot \mathbf{x},\ 0 \leq \mathbf{x} \leq \mathbf{1}$
	$\epsilon \in \mathrm{Eff}_1(A) \equiv \mathrm{Trn}_1(A \to I)$	$\epsilon = \cdot \mathbf{1}$

Acknowledgements The authors acknowledge helpful conversations and encouragement by Don Hoffman, and interesting discussions with Chris Fields. Giacomo Mauro D'Ariano acknowledges interesting discussions with Ramon Guevarra Erra. This work has been sponsored by the Elvia and Federico Faggin foundation through the Silicon Valley Community Foundation, Grant 2020-214365 *The observer: an operational theoretical approach.* For oral sources see also the Oxford podcast http://podcasts.ox.ac.uk/mauro-dariano-awareness-operational-theoretical-approach.

References

1. Chalmers, D. J. (2010). *The character of consciousness.* Oxford University Press.
2. Dennett, D. (2017). *Consciousness explained.* Little, Brown.
3. Wheeler, J. (1989). Information, physics, quantum: The search for link. In *Proceedings of 3rd International Symposium on Foundations of Quantum Mechanics*, Tokyo.
4. Feynman, R. (1982). *International Journal of Theoretical Physics, 21,* 467–488.
5. Hey, A. J. (Ed.). (1998). *Feynman and computation.* Perseus Books.
6. Finkelstein, D. (1996). *Quantum relativity: A synthesis of the ideas of Einstein and Heisenberg.* Springer.
7. D'Ariano, G. M. (2017). *International Journal of Theoretical Physics, 56,* 97–128.
8. Chiribella, G., D'Ariano, G. M., & Perinotti, P. (2011). *Physical Review A, 84,* 012311.
9. Chiribella, G., D'Ariano, G. M., & Perinotti, P. (2010). *Physical Review A, 81,* 062348.
10. D'Ariano, G. M., Chiribella, G., & Perinotti, P. (2017). *Quantum theory from first principles.* Cambridge University Press.
11. D'Ariano, G. M., & Perinotti, P. (2014). *Physical Review A, 90*(6), 062106.

12. Bisio, A., D'Ariano, G. M., & Perinotti, P. (2016). *Annals of Physics, 368*, 177–190.
13. D'Ariano, G. M. (2018). *Philosophical Transactions of the Royal Society A: Mathematical, Physical and Engineering Sciences, 376*, 20170224.
14. Susskind, L. (1995). *Journal of Mathematical Physics, 36*, 6377.
15. Bousso R 2003 *Phys. Rev. Lett.* **90**(12) 121302
16. Fuchs, C. A. (2002). *arXiv quant-ph/0205039.*
17. Nielsen, M. A., & Chuang, I. L. (1997). *Physical Review Letters, 79*, 321–324.
18. Shannon, C. E. (1948). *Bell System Technical Journal, 27*, 379–423.
19. Chiribella, G., D'Ariano, G. M., & Perinotti, P. (2016). *Quantum from principles* (pp. 171–221). Springer.
20. Coecke, B., & Lal, R. (2011). *QPL, 2011*, 67.
21. Tull, S. (2020). *Logical Methods in Computer Science, 16*(1), 1860–5974.
22. Coecke, B., & Lal, R. (2012). *Foundations of Physics, 43*, 458–501.
23. Coecke, B., Fritz, T., & Spekkens, R. W. (2016). *Information and Computation, 250*, 59–86.
24. Abramsky S and Heunen C 2016 *Operational theories and categorical quantum mechanics* (Cambridge University Press) pp 88--122 Lecture Notes in Logic
25. Coecke, B., & Kissinger, A. (2018). *Categorical quantum mechanics I: Causal quantum processes.* Oxford University Press.
26. Popescu, S., & Rohrlich, D. (1992). *Physics Letters A, 166*, 293–297.
27. Loorits, K. (2014). *Frontiers in Psychology, 5.*
28. Tosini, A., D'Ariano, G. M., & Perinotti, P. (2019) *Quantum* (in press).
29. Raichle, M. E. (2010). *Trends in Cognitive Sciences, 14*, 180–190.
30. Anderson, C. H., Essen, D. C. V., & Olshausen, B. A. (2005). Directed visual attention and the dynamic control

of information flow. *Neurobiology of attention* (pp. 11–17). Elsevier.
31. Nørretranders, T. (1998). *The user illusion: Cutting consciousness down to size*. Viking.
32. Olshausen, B. A., & Field, D. J. (2005). *Neural Computation, 17,* 1665–1699.
33. Holevo, A. (1973). Bounds for the quantity of information transmitted by a quantum communication channel. *Problems of Information Transmission, 9,* 177–183.
34. D'Ariano, G. M. (2003). *Fortschritte der Physik: Progress of Physics, 51,* 318–330.
35. Chiribella, G., D'Ariano, G. M., & Schlingemann, D. (2007). *Physical Review Letters, 98,* 190403.
36. D'Ariano, G. M. (2000). *Fortschritte der Physik, 48,* 579–588.
37. Fuchs, C., Hoang, M., & Stacey, B. (2017). *Axioms, 6,* 21.
38. Plesch, M., & Brukner, C. (2011). *Physical Review A, 83*(3), 032302.
39. Hayashi, A., Hashimoto, T., & Horibe, M. (2005). *Physical Review A, 72*(3), 032325.
40. Dehaene, S. (2014). *Consciousness and the brain: Deciphering how the brain codes our thoughts.* Penguin Books
41. Chalmers, D. (2016). The combination problem for panpsychism. In G. Brüntrup & L. Jaskolla (Eds.), *Panpsychism*. Oxford University Press.
42. Bennett, C. H., Brassard, G., Crépeau, C., Jozsa, R., Peres, A., & Wootters, W. K. (1993). *Physical Review Letters, 70,* 1895.
43. Benedek, G., & Caglioti, G. (2019) Graphics and quantum mechanics—The necker cube as a quantum-like two-level system. In L. Cocchiarella (Ed.), *ICGG 2018—Proceedings of the 18th International Conference on Geometry and Graphics* (pp. 161–172). Springer International Publishing.
44. (1993). *Magic eye: A new way of looking at the world: 3D illusions.* Andrews and McMeel.
45. Jahn, R. (2007). *EXPLORE, 3,* 307–310.
46. Tononi, G. (2008). *The Biological Bulletin, 215,* 216–242.
47. Mac Lane, S. (1978). *Categories for the working mathematician* (Vol. 5). Springer Science & Business Media.

48. Selinger, P. (2011). A survey of graphical languages for monoidal categories. *New structures for physics* (pp. 289–355). Springer
49. Coecke, B. (2006). Introducing categories to the practicing physicist. *What is category theory? Advanced studies in mathematics and logic* (Vol. 30, pp. 289–355) Polimetrica Publishing.
50. Coecke, B. (2006). *Advanced Studies in Mathematics and Logic, 30*, 45.
51. Coecke, B., Moore, D., & Wilce, A. (2000). *Current research in operational quantum logic: Algebras, categories, languages* (Vol. 111). Springer.
52. Coecke, B. (2005). In G. Adenier, A. Khrennikov, & T. M. Nieuwenhuizen (Eds.). *Quantum theory: Reconsideration of foundations-3*. AIP Conference Proceedings (Vol. 810, p. 81). American Institute of Physics.
53. Coecke, B. (2010). *Contemporary Physics, 51*, 59–83.
54. Hardy, L. (2001). *arXiv quant-ph/0101012*.
55. Hardy, L. (2007). *Journal of Physics A: Mathematical and Theoretical, 40*, 3081.
56. Hardy, L. (2013). *Mathematical Structures in Computer Science, 23*, 399–440.
57. D'Ariano, G. M., Manessi, F., Perinotti, P., & Tosini, A. (2014). *EPL (Europhysics Letters), 107*, 20009.
58. D'Ariano, G. M., Erba, M., & Perinotti, P. (2020). *Physical Review A, 101*(4), 042118.
59. D'Ariano, G. M., Erba, M., & Perinotti, P. (2020). *Physical Review A, 102*(5), 052216.
60. Barrett, J. (2007). *Physical Review A, 75*, 032304.
61. D'Ariano, G. M. (2020). *Foundations of Physics, 50*, 1921–1933.

Open Access This chapter is licensed under the terms of the Creative Commons Attribution 4.0 International License (http://creativecommons.org/licenses/by/4.0/), which permits use, sharing, adaptation, distribution and reproduction in any medium or format, as long as you give appropriate credit to the original author(s) and the source, provide a link to the Creative Commons license and indicate if changes were made.

The images or other third party material in this chapter are included in the chapter's Creative Commons license, unless indicated otherwise in a credit line to the material. If material is not included in the chapter's Creative Commons license and your intended use is not permitted by statutory regulation or exceeds the permitted use, you will need to obtain permission directly from the copyright holder.

CPSIA information can be obtained
at www.ICGtesting.com
Printed in the USA
LVHW061411010623
748615LV00001B/1